97 Advances in Polymer Science

Synthesis/Mechanism/ Polymer Drugs

With contributions by
M. Akashi, H. K. Hall, C. Lee, T. Li,
W. Kamińska, P. Penczek, A. D. Pomogailo
K. Takemoto, I. E. Uflyand

With 35 Figures and 32 Tables

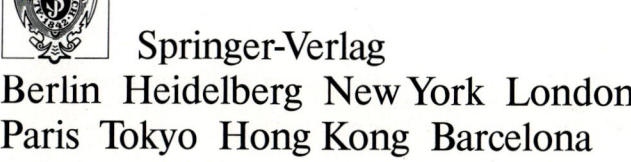

Springer-Verlag
Berlin Heidelberg New York London
Paris Tokyo Hong Kong Barcelona

ISBN-3-540-52834-2 Springer-Verlag Berlin Heidelberg New York
ISBN-0-387-52834-2 Springer-Verlag New York Berlin Heidelberg

Library of Congress Catalog Card Number 61-642

This work is subject to copyright. All rights are reserved, whether the whole or part of the material is concerned, specifically the rights of translation, reprinting, reuse of illustrations, recitation, broadcasting, reproduction on microfilms or in other ways, and storage in data banks. Duplication of this publication or parts thereof is only permitted under the provisions of the German Copyright Law of September 9, 1965, in its current version, and a copyright fee must always be paid.

© Springer-Verlag Berlin Heidelberg 1990
Printed in Germany

The use of registered names, trademarks, etc. in this publication does not imply, even in the absence of a specific statement, that such names are exempt from the relevant protective laws and regulations and therefore free for general use.

Typesetting: Th. Müntzer, Bad Langensalza; Printing: Heenemann, Berlin; Bookbinding: Lüderitz & Bauer, Berlin
2152/3020-543210 — Printed on acid-free paper

Editors

Prof. Akihiro Abe, Tokyo Institute of Technology, Faculty of Engineering, Department of Polymer Chemistry, O-okayama, Meguro-ku, Tokyo 152, Japan
Prof. Henri Benoit, CNRS, Centre de Recherches sur les Macromolecules, 6, rue Boussingault, 67083 Strasbourg Cedex, France
Prof. Hans-Joachim Cantow, Institut für Makromolekulare Chemie der Universität, Stefan-Meier-Str. 31, 7800 Freiburg i. Br., FRG
Prof. Paolo Corradini, Università di Napoli, Dipartimento di Chimica, Via Mezzocannone 4, 80134 Napoli, Italy
Prof. Karel Dušek, Institute of Macromolecular Chemistry, Czechoslovak Academy of Sciences, 16206 Prague 616, ČSSR
Prof. Sam Edwards, University of Cambridge, Department of Physics, Cavendish Laboratory, Madingley Road, Cambridge CB3 OHE, England
Prof. Hiroshi Fujita, 35 Shimotakedono-cho, Shichiku, Kita-ku, Kyoto 603, Japan
Prof. Dr. Hartwig Höcker, Deutsches Wollforschungs-Institut e. V. an der Technischen Hochschule Aachen, Veltmanplatz 8, 5100 Aachen, FRG
Prof. Hans-Henning Kausch, Laboratoire de Polymères, Ecole Polytechnique Fédérale de Lausanne, 32, ch. de Bellerive, 1007 Lausanne, Switzerland
Prof. Joseph P. Kennedy, Institute of Polymer Science. The University of Akron, Akron, Ohio 44325, U.S.A.
Prof. Anthony Ledwith, Pilkington Brothers plc. R & D Laboratories, Lathom Ormskirk, Lancashire L40 5UF, U.K.
Prof. Seizo Okamura, No. 24, Minamigoshi-Machi Okazaki, Sakyo-Ku, Kyoto 606, Japan
Prof. Charles G. Overberger, Department of Chemistry. The University of Michigan, Ann Arbor, Michigan 48109, U.S.A.
Prof. Helmut Ringsdorf, Institut für Organische Chemie, Johannes-Gutenberg-Universität, J.-J.-Becher Weg 18–20, 6500 Mainz, FRG
Prof. Takeo Saegusa, Department of Synthetic Chemistry, Faculty of Engineering, Kyoto University, Yoshida, Kyoto, Japan
Prof. J. C. Salamone, University of Lowell, Department of Chemistry, College of Pure and Applied Science, One University Avenue, Lowell, MA 01854, U.S.A.
Prof. John L. Schrag, University of Wisconsin, Department of Chemistry, 1101 University Avenue. Madison, Wisconsin 53706. U.S.A.
Prof. John K. Stille, Department of Chemistry, Colorado State University, Fort Collins, Colorado 80523, U.S.A.
Prof. Dr. G. Wegner, Max-Planck-Institut für Polymerforschung, Akkermannweg 10, Postfach 3148, 6500 Mainz, FRG

Table of Contents

The Role of Tetramethylene Diradicals in Photo-Induced "Charge-Transfer" Cycloadditions and Copolymerization
T. Li, C. Lee, H. K. Hall 1

Polyfunctional Cyanate Monomers as Components of Polymer Systems
P. Penczek, W. Kamińska 41

Polymers Containing Metallochelate Units
A. D. Pomogailo, I. E. Uflyand 61

New Aspects of Polymer Drugs
M. Akashi, K. Takemoto 107

Author Index Volumes 1–97 147

Subject Index 163

The Role of Tetramethylene Diradicals in Photo-Induced "Charge-Transfer" Cycloadditions and Copolymerization

Tong Li*, Cherylyn Lee**, H. K. Hall, Jr.

C. S. Marvel Laboratories, Department of Chemistry, University of Arizona, Tucson, AZ 85721, USA

A mechanism for the photo-induced copolymerization of donor monomers with acceptor monomers is proposed based on a combination of the results from organic chemistry and the results from thermal polymerizations and photopolymerizations. Under certain conditions, tetramethylene diradicals, the proposed key intermediates, are formed from the exciplex between the monomer pairs and undergo cyclization and initiate copolymerization.

1 Brief Review of Photoreactions of Donor/Acceptor Pairs 3
  1.1 Neither Donor nor Acceptor is Polymerizable 3
  1.2 One of the Components is Homopolymerizable 5
  1.3 D Copolymerizable with A 7
2 Exciplexes in Photopolymerizations and Cycloadditions 9
  2.1 Exciplex and Excited CT Complex —
    Two Routes of Photoactivation 10
    2.1.1 CT Complex Excitation 10
    2.1.2 Monomer Excitation 11
    2.1.3 Effect of D/A Strength on Polymerization 11
  2.2 Further Discussion of Exciplexes 13
    2.2.1 Emitting and Non-Emitting Exciplexes 13
    2.2.2 Polarity of Exciplex 14
    2.2.3 Triplet and Singlet Exciplexes 14
    2.2.4 Triplex 15
  2.3 Internal Chemistry of Exciplexes 16
    2.3.1 Formation of Two C—C Bond: Concerted Cycloaddition . . 16
    2.3.2 Formation of One C—C Bond: Tetramethylene Intermediates 17
    2.3.3 No C—C Bond Formation: Ion-Radical Formation 17
    2.3.4 Loss of One Bond: Proton Transfer Reaction 20
3 Tetramethylenes in Spontaneous Thermal Reactions
  of Donor/Acceptor Olefin Pairs 21

* Currently visiting scholar from Department of Material Science, Fudan University, Peoples Republic of China.
** Current address: Hoechst-Celanese Research Co., 86 Morris Ave., Summit, NJ 07901

4 Tetramethylene Diradicals as Key Intermediates in Photoreactions of Donor/Acceptor Olefins 23
   4.1 Tetramethylene Diradicals in Photochemistry 23
   4.2 Tetramethylene Diradicals as Intermediates in Cycloadditions of Donor/Acceptor Olefins 26
   4.3 Tetramethylene Diradicals as Intermediates in Copolymerization of Donor/Acceptor Olefins 34

5 Conclusion 35

6 References 36

1 Brief Review of Photoreactions of Donor/Acceptor Pairs

In the last twenty years there has been increasing interest in "charge-transfer polymerizations". The essential character of such polymerizations is the interaction between an electron-donor (D) and an electron-acceptor (A) monomer in the initiation and/or propagation processes.

Charge-transfer polymerizations fall into two classes: in thermal polymerizations charge-transfer (CT) interaction takes place in the ground state, while in photopolymerizations the CT interaction plays an essential role in the excited state. Photopolymerizations have attracted great attention due to two reasons. First, in the excited state organic compounds become stronger donors or acceptors. Charge-transfer occurs even for those combinations which have no interaction in the ground state. Therefore, a wider range of donor/acceptor pairs is available. Second, the rapidly developing organic photochemistry and photophysics provide powerful means, both theoretically and practically (such as flash photolysis and laser photolysis), for the understanding of the mechanism.

An electron-donor molecule possesses a high electron density due to the presence of delocalized electrons or non-bonding electrons, such as those associated with nitrogen, oxygen, and sulfur. An electron-acceptor molecule, on the other hand, is able to accept electron density and stabilize the additional charge through resonance or induction. According to the polymerizability of the components, photo-induced reactions of donor/acceptor pairs are divided into three types: both are non-polymerizable (Type I), one is polymerizable (Type II), or both are polymerizable (Type III).

1.1 Neither Donor nor Acceptor is Polymerizable

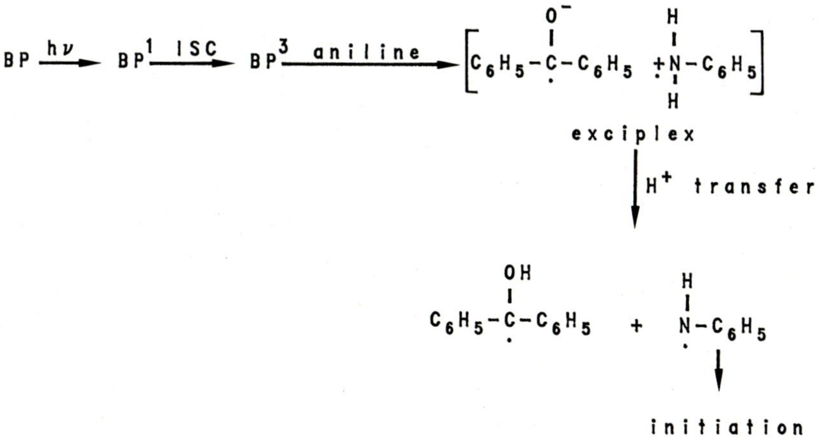

$$BP \xrightarrow{h\nu} BP^1 \xrightarrow{ISC} BP^3 \xrightarrow{aniline} \left[C_6H_5-\overset{O^-}{\underset{\cdot}{C}}-C_6H_5 \quad \overset{H}{\underset{H}{\overset{+}{N}}}-C_6H_5 \right]$$

$$\text{exciplex}$$

$$\downarrow H^+ \text{ transfer} \qquad (1)$$

$$C_6H_5-\overset{OH}{\underset{\cdot}{C}}-C_6H_5 \quad + \quad \overset{H}{\underset{\cdot}{N}}-C_6H_5$$

$$\downarrow$$

$$\text{initiation}$$

A. *Bond is Not Formed.* The electron donor/acceptor combinations afford initiating systems for vinyl polymerizations. The notable combinations consist of a sensitizer, such as an aromatic ketone, and an amine, which acts as both the electron-donor and the hydrogen-donor. A typical system is benzophenone (BP)/aniline, which forms an exciplex under irradiation and then undergoes proton-transfer, producing two radicals (see(1)).

Of the two kinds of radicals, the amino radicals are responsible for the initiation, while the semipinacol radical plays a role in the primary radical termination [1].

B. *Bond-Formation is Possible*. Cycloaddition is the most popular reaction studied. For example, mixed cyclodimers can be formed by irradiation of 1:1 mixtures of two different anthracene derivatives [2, 3]:

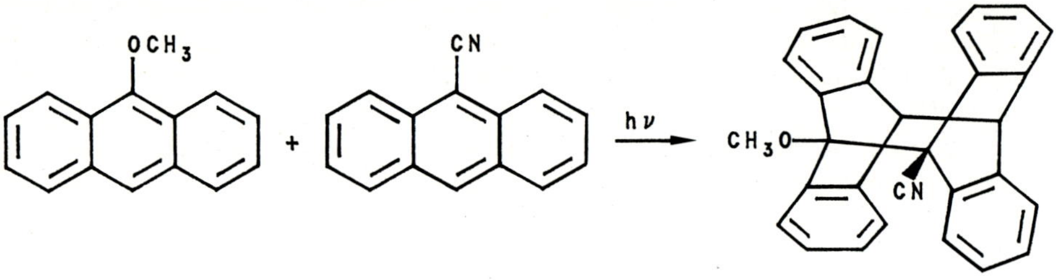

(2)

In some other reactions, only one bond is formed. Naphthalene reacts with pyrrole on irradiation, forming dihydronaphthalenes [4].

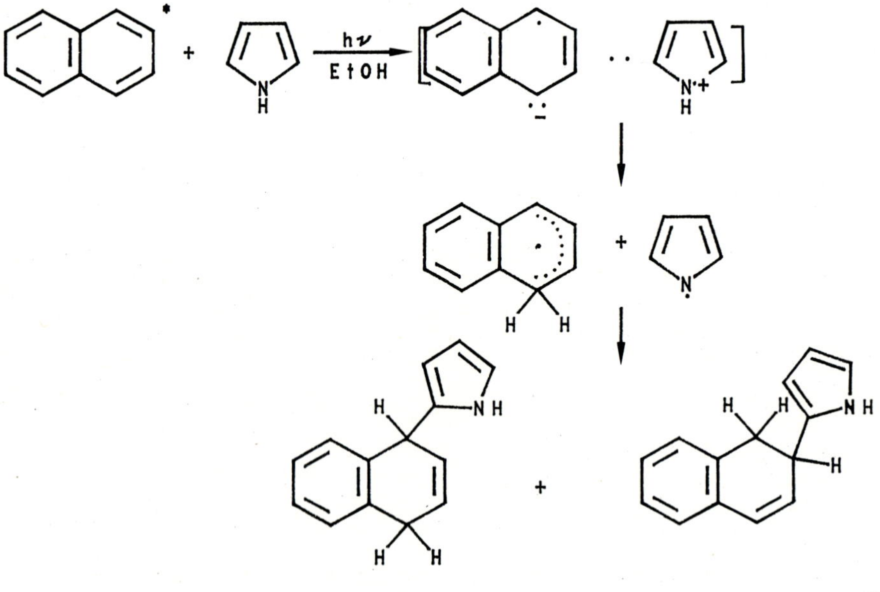

(3)

1.2 One of the Components is Homopolymerizable

A. *D is Homopolymerizable, A is Not; Cationic Homopolymerization and/or "Homo"-Cycloaddition of D*

Notable work in the area of photopolymerizations of donor monomers initiated by acceptor initiators was done by Shirota [5–7] in his study of polymerization of *N*-vinylcarbazole (VCZ) and by Hayashi and Irie [8] on the polymerization of α-methylstyrene (α-MSt). The initiation mechanism was proposed to proceed via the charge-transfer complex between VCZ (or α-MSt) and the acceptor, which then yields two kinds of ion-radicals $D^{\ddagger}$ and $A^{\overline{\cdot}}$.

The $D^{\ddagger}$ ion-radicals may either directly initiate cationic polymerization or undergo subsequent reactions to produce the initiating species. For the VCZ/A system, the primary process of photoreaction was proposed as follows [4]:

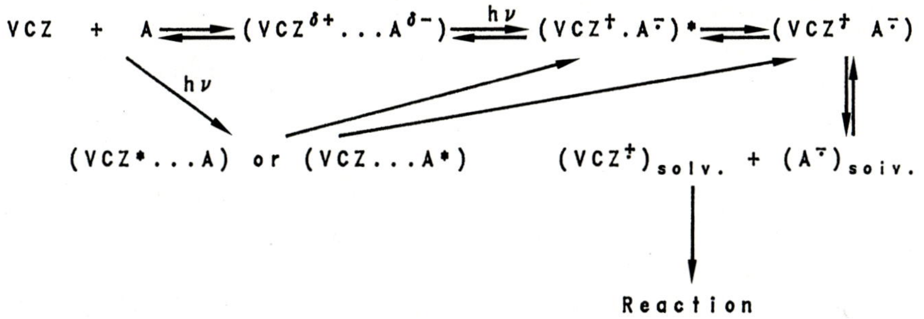

$$VCZ + A \rightleftharpoons (VCZ^{\delta+}\ldots A^{\delta-}) \xrightarrow{h\nu} (VCZ^{\ddagger}\cdot A^{\overline{\cdot}})^* \rightleftharpoons (VCZ^{\ddagger} \; A^{\overline{\cdot}})$$

with $h\nu$ pathway to $(VCZ^*\ldots A)$ or $(VCZ\ldots A^*)$, and dissociation to $(VCZ^{\ddagger})_{solv.} + (A^{\overline{\cdot}})_{solv.}$ followed by Reaction.

(4)

Reaction (5): starting from $\overset{\sim N \sim}{\underset{CH=CH_2}{|}}$ → (with A) $\overset{\sim \overset{+}{N} \sim}{\underset{CH=CH_2}{|}}$ → (with VCZ) the dimer cation → (e) the neutral dimer, leading to:

- cationic polymerization in benzene → $\overset{\sim N \sim}{\underset{\{CH-CH_2\}_n}{|}}$
- cyclodimerization in acetonitrile
- radical polymerization in HMPA (via H^+ loss giving the allylic radical species)

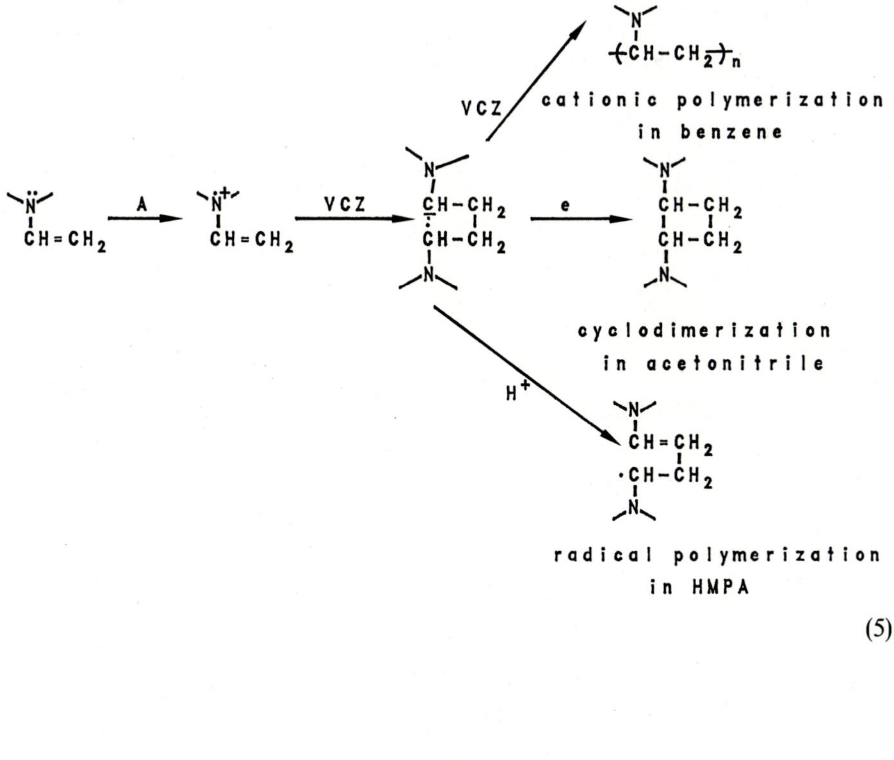

(5)

Solvent polarity influences the dissociation of the intimate ion pair, while the fate of the ion-radicals depends mainly on the solvent basicity. In non-basic solvents (benzene or dichloromethane), cationic polymerization of VCZ takes place. In moderately basic solvents (acetone or acetonitrile), VCZ cyclodimerizes. Radical polymerization occurs along with cycloaddition in strongly basic solvents, such as DMF and DMSO, while only radical polymerization takes place in the extremely basic solvent hexamethylphosphoric triamide (HMPA).

For α-MSt/A system, the mechanistic scheme was proposed as follows:

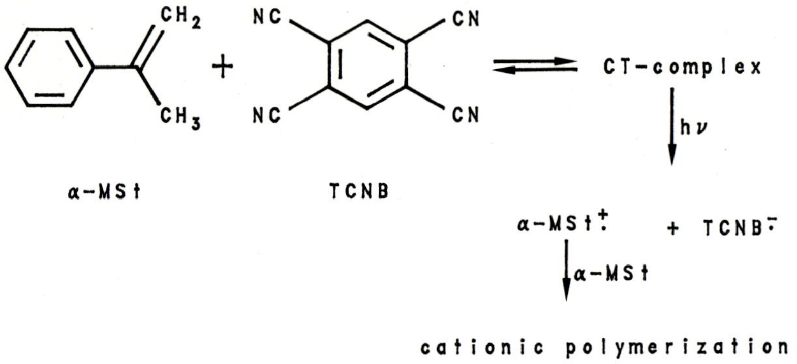

$$\alpha\text{-MSt} + \text{TCNB} \rightleftharpoons \text{CT-complex} \xrightarrow{h\nu} \alpha\text{-MSt}^{+\cdot} + \text{TCNB}^{-\cdot} \xrightarrow{\alpha\text{-MSt}} \text{cationic polymerization} \tag{6}$$

In this case the monomer ion radical would directly initiate the polymerization.

B. *A is Homopolymerizable, D is Not; Radical Homopolymerization of A*

Feng, Cao, and Li [9] studied the photopolymerization of acceptor monomers initiated by donor initiators, especially by aromatic amines. The proposed mechanism involved a charge-transfer interaction, followed by proton-transfer to produce two kinds of free radicals. The acrylonitrile (AN)/*N,N*-dimethyltoluidine (DMT) system is shown as an example:

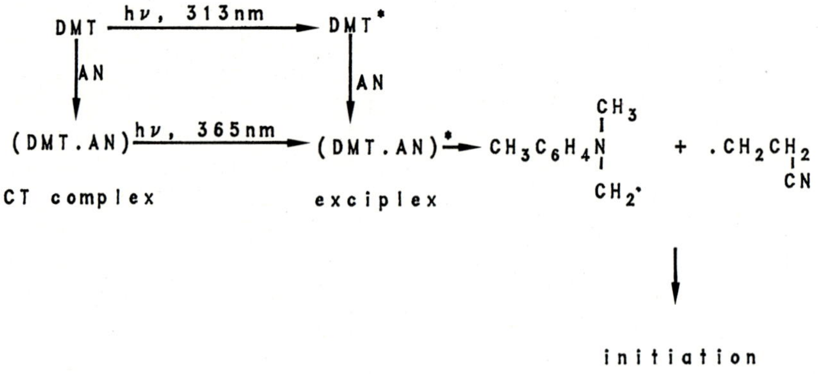

$$\begin{array}{c} \text{DMT} \xrightarrow{h\nu,\ 313\text{nm}} \text{DMT}^* \\ \downarrow \text{AN} \qquad\qquad\qquad \downarrow \text{AN} \\ (\text{DMT}\cdot\text{AN}) \xrightarrow{h\nu,\ 365\text{nm}} (\text{DMT}\cdot\text{AN})^* \longrightarrow \text{CH}_3\text{C}_6\text{H}_4\overset{\text{CH}_3}{\underset{\text{CH}_2^\bullet}{\text{N}}} + {}^\bullet\text{CH}_2\text{CH}_2\text{CN} \\ \text{CT complex} \qquad\qquad \text{exciplex} \\ \qquad\qquad\qquad\qquad\qquad\qquad \downarrow \\ \qquad\qquad\qquad\qquad\qquad\qquad \text{initiation} \end{array} \tag{7}$$

Naphthalene and certain of its alkoxy and alkyl derivatives react with AN on irradiation, giving cyclobutane adducts [10, 11]. An example is given below:

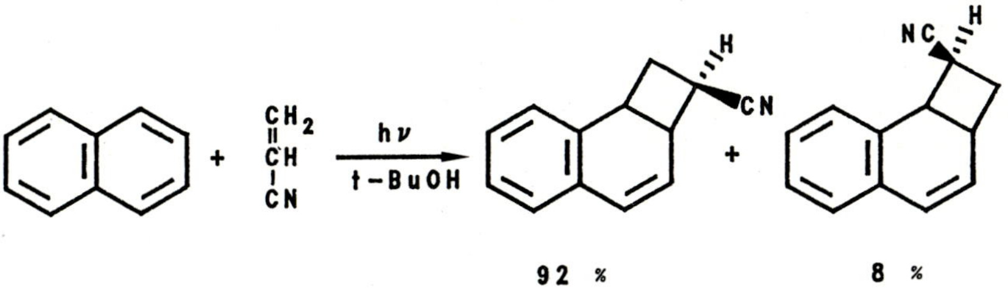

$$\text{naphthalene} + \underset{\underset{CN}{|}}{\overset{CH_2}{\underset{\|}{CH}}} \xrightarrow[t-BuOH]{h\nu} \text{adduct (92\%)} + \text{adduct (8\%)} \quad (8)$$

Also, naphthalene was reported to photoinitiate the polymerization of AN [12, 13].

1.3 D is Copolymerizable with A

The photo-induced charge-transfer copolymerization entity consists of a donor monomer and an acceptor monomer, without initiator. Reported combinations usually consist of an aryl vinyl monomer as the donor component and a 1,2-disubstituted vinyl monomer as the acceptor component.

One of the important considerations in the choice of comonomers is to meet the light absorption requirement. To perform photopolymerization, the system must absorb light in the region where the excitation source has a significant output. Under usual photopolymerization conditions (using Pyrex vessels), light <300 nm from the high-pressure mercury lamp is cut off, leaving only light at 313 nm and 365 nm. Aryl vinyl monomers, due to their large conjugated substituents, usually absorb at much longer wavelengths than aliphatic monomers. Their absorption bands can either match the light from the mercury lamp so that it is possible to directly excite these monomers, or they can be broadened to even longer wavelengths due to formation of CT complexes with moderately strong acceptors, so that these CT complexes can be selectively excited by choosing suitable wavelengths.

1,2-Disubstituted vinyl comonomers are desirable because they have moderate electron-accepting ability and do not readily undergo radical homopolymerization. Among these, maleic anhydride (MAn) is most widely used, for example, in VCZ/MAn [6], St/MAn [14–17], butadiene/MAn [18], isoprene/MAn [18], cyclohexene/MAn [19, 20], stilbene/MAn [19], butyl vinyl ether/MAn [21], vinyl acetate (VAc)/MAn [22], t-butylstyrene/MAn [23] and 2-vinylnaphthalene(2-VN)/MAn [24] systems. MAn, with its absorption maximum at 260 nm, has an absorption band which extends to 310 nm [25] and is extended to even longer wavelengths by forming a CT complex with a donor. Therefore, even aliphatic monomers can serve as partners to form photocopolymerizable systems with MAn, such as the VAc/MAn system. All these systems produced alternating copolymers by a radical mechanism.

Another acceptor is fumaronitrile (FN). In spite of its lack of absorption above 300 nm, the partners or the CT complexes may absorb light and thus bring about polymerization. In studying 2-VN/FN system, Shirota obtained not only 1:1 alternating copolymer but also several cycloaddition products [26]:

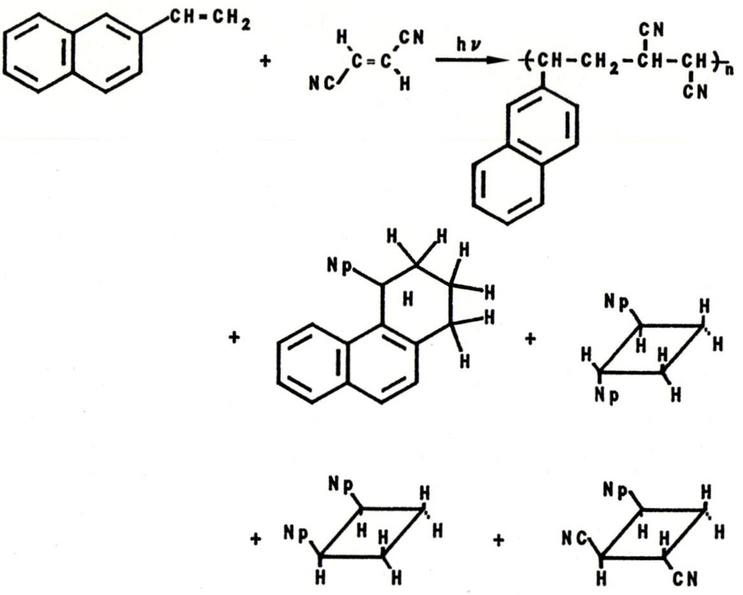

$$\text{(9)}$$

(Np = 2-naphthyl)

The photocopolymerization of the *N*-vinylindole/FN system yielded an alternating copolymer with partial addition of FN at the electron-rich 3-position of the indole ring [27]:

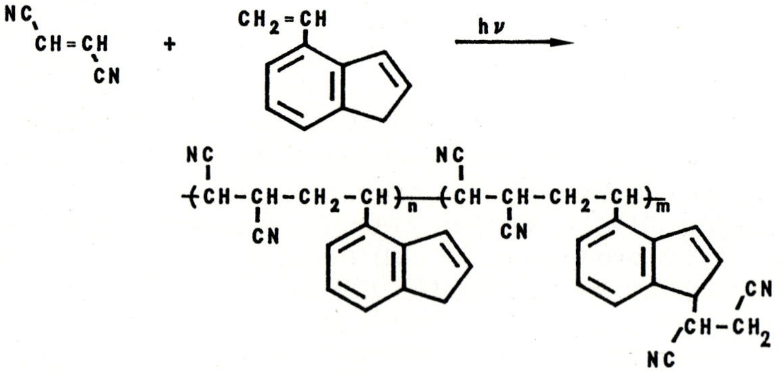

$$\text{(10)}$$

Acrylonitrile (AN), a weak electron-accepting but very polar monomer, was also reported to photocopolymerize with donor monomers, such as St [28], isobutyl vinyl ether [29], or butadiene [30].

Generally, the photo-induced charge-transfer copolymerization follows a radical mechanism and produces alternating copolymers. Accompanying cyclobutane formation is usually observed, which provides information for understanding the mechanism.

For the strong donor monomer VCZ, the photoreaction depends on the solvent basicity and the molar ratio of the donor and the acceptor. In strongly basic solvents such as dimethyl formamide (DMF), the radical homopolymerization of VCZ occurs in the presence of catalytic amounts of FN or diethyl fumarate (DEF), but it is replaced by radical copolymerization in an equimolar amount of the monomers. The cationic homopolymerization of VCZ, which proceeds in less basic solvents, e.g., benzene, and the cyclodimerization of VCZ, which proceeds in moderately basic solvents, e.g., acetone, is accompanied by the radical copolymerization of VCZ with FN or DEF [6].

In studying the photoreaction of donor/acceptor pairs, photochemists work in two different ways. Organic chemists select carefully non-polymerizable pairs to avoid polymerization and prefer to use dilute solutions to favor intramolecular reactions. They add inhibitors and discard polymeric products before isolation of the "product". Polymer chemists, on the other hand, are deliberately making large organic molecules, they use polymerizable olefins, they isolate the product by precipitating the reaction mixture and filtering; small organic molecules are thus conveniently removed.

Both polymer chemists and organic chemists have obtained much information about these reactions, meanwhile, considerable information has been lost by their techniques. Small-molecule formation accompanying polymerization can give us important clues about the nature of the initiating species. In this review, we will attempt to integrate the published results from both sides.

2 Exciplex in Photopolymerizations and Cycloadditions

Photoreactions of donor/acceptor systems are conventionally initiated by using one of the following three routes:
(1) direct irradiation of monomers (usually the donor);
(2) select irradiation of CT complexes;
(3) irradiation of a triplet sensitizer and transfer of the energy to the monomer.

For photocopolymerization certain initiators may be added and irradiated giving rise to either free radical or cationic species via homolytic bond cleavage, hydrogen abstraction or electron transfer. This kind of photoreaction, however, is not discussed in this review.

As many photochemists have pointed out, photocopolymerizations and photocycloadditions usually involve charge-transfer complexes as intermediates, either in the ground state or in the excited state. The charge-transfer complexes that are formed only in the electronically excited state of either donor or acceptor molecules are termed exciplexes, which are stable in the electronically excited state, but dissociate in the ground state. Exciplexes are so common that we have to give a brief introduction about their role in photocycloadditions and photopolymerizations.

2.1 Exciplex and Excited CT Complex — Two Routes of Photoactivation

2.1.1 CT Complex Excitation

A charge-transfer complex (CT complex or CTC) is involved in all thermal charge-transfer polymerizations and in most of the photo-induced charge-transfer copolymerizations. Mixing of a donor (D), such as VCZ, with an acceptor, such as tetracyanoethylene (TCNE), immediately produces color [31] due to the formation of a CT complex:

$$D + A \rightleftarrows CT\ complex.$$

It can also be called an electron donor-acceptor (EDA) complex. The electron-rich compound donates electron density to the electron-poor compound, without full electron transfer or bond formation, generating a weakly bound complex which does not display an ESR signal. So we should clearly distinguish "charge-transfer" from "electron transfer". It might be better to think of "partial electron transfer" and "full electron transfer". Formation of such complexes increases their mutual reactivity by bringing the reactants closer together, allowing orbital overlap for subsequent bond formation or electron-transfer.

Because the complex exists only in solution and is not isolable as a discrete moiety, it may be characterized only by spectroscopic methods, such as NMR and ultraviolet-visible (UV-VIS) spectroscopy. The most important character of a CT complex is the new absorption band in UV spectra, which is attributed to a charge-transfer transition. Strong donor and strong acceptor systems show a new, very broad absorption band in UV-VIS spectra, usually in the visible region, developing a bright or deep color. In contrast, weak donors and acceptors form contact charge-transfer (CCT) complexes, which do not show an isolated absorption peak and whose absorption bands overlap with the donor or acceptor absorption bands with only slight broadening. For a given donor, the absorption character of the CT complexes depends on the acceptor strength.

Equilibrium constants K may be determined from UV-VIS data, using the Benesi-Hildebrand method, or by NMR spectroscopy, using the Hanna-Asbaugh relationship. The relative strength of donor-acceptor interactions may be assessed by two characteristics of the complex: the position of the band in the spectrum and the equilibrium constant. A strong interaction is expected to have a greater K value and an absorption peak shifted to longer wavelengths. UV data of some CT complexes are collected in Table 1 [23].

Table 1. UV data of CT complexes

p-substituent of styrene	Acceptor		
	Dimethyl fumarate	FN	MAn
$-O(CH_2)_2CH(CH_3)_2$	324[a] (1.69[b])	360 (2.56)	340 (0.480)
$-OCH_2CH_2Cl$	329 (0.838)	350 (2.70)	335 (0.452)
$-OCH_3$			335 (0.525)
$-C(CH_3)_3$	312 (0.688)	350 (1.87)	330 (0.338)
$-CH_3$	312 (0.938)	350 (2.27)	330 (0.424)

[a] λ max$^{(nm)}$;
[b] Absorbance measured in $CHCl_3$ at room temperature (26 °C); [D] = [A] = 0.2 mol/L; MAn complexes measured in 1-mm cell; others measured in 10-mm cell.

As to chemical reactivity, whether or not CT complexes react with external reagents is a vigorously debated topic. However, intramolecular reactions of the components with each other are firmly established.

As mentioned above, in photoreactions, the intermediates are formed through the excitation of one partner or the excitation of CT complex. In the former case, the excited molecule interacts with the partner to form the exciplex, while the latter produces "excited EDA complex" or in short, "excited complex". Actually, some studies have shown that exciplexes and excited complexes are identical species [32–35]: 1) They have the same spectral character. For example, when 0.12 M *trans*-stilbene and 0.43 M fumaronitrile are excited in the charge-transfer absorption band (360 nm), both the spectral distribution and lifetime of this emission are identical with those obtained for their exciplex [36]. 2) They undergo the same follow-up reactions. Lewis [35] reported a similar cycloaddition quantum yield of the two above processes in the photocycloaddition of stilbene with dimethyl fumarate. Accordingly, here we will use them interchangeably.

2.1.2 Monomer Excitation

Even if no charge-transfer interaction exists in the ground state for weak donor/acceptor pairs, charge-transfer complexes can be formed under irradiation, through the interaction of a locally excited molecule with a suitable ground state donor or acceptor molecule. This is because both the electron-donating and the electron-accepting ability of organic molecules are greatly enhanced in their excited state.

2.1.3 Effect of D/A Strength on Polymerization

Not much attention has been paid to the two different kinds of excitation in photo-induced copolymerizations. A number of authors do not mention which species was excited, one of the partners or the CT complex. However, there are investigators who have made sure to excite the CT complex only by selecting the wavelength of the incident light [23]. Neither St nor FN absorbs 365-nm light, while their CT complex does. Therefore, the excited CT complex plays an essential role in the initiation when 365-nm light is used [37]. St/AN, on the other hand, may serve as an example in which only local excitation of St plays a role [38]. No CT complex was observed in UV-VIS spectra for this system, but the St molecule can absorb the light, and then form an excimer with another St molecule or form an exciplex with AN molecule, which can further collapse to an initiating species.

The following conclusions can be drawn about the effect of D/A strength on polymerization:

(1) Since the excited charge-transfer complexes are photoactivated, weaker donor-acceptor combinations can be used than those in thermal polymerizations, including not only strong D/strong A, but also strong D/weak A and weak D/strong A.

Some D/A combinations do not interact strongly enough to bring about thermal copolymerization, but readily yield copolymers under irradiation, for example, p-t-butylstyrene/MAn [23]. Some undergo very slow thermal polymerization, while the photopolymerization is much faster, such as St/FN system. Table 2 [37] shows the results of thermal (80 °C) and photocopolymerizations of some monomer pairs. The acceptor strength decreases in the following order:

$$\text{diethyl fumarate (DEF)} < \text{diethyl maleate (DEM)} < \text{FN} < \text{MAn}$$
$$(1.25) \qquad\qquad (1.52) \qquad\qquad (1.96)\ (2.25)$$

The numbers in the brackets show the e values of Price-Alfrey equation. The donor strength decreases in the following order:

$$\text{VAc} < \text{2-VN} < \text{St} < \alpha\text{-MeSt}$$
$$(-0.02)\quad (-0.38)\quad (-0.80)\quad (-1.27)$$

Spontaneous thermal copolymerization is usually observed in strong donor (D)/strong acceptor (A) pairs, for example α-MSt/MAn, because only in these pairs is the charge-transfer interaction strong enough to produce initiating radicals. On the other hand, photo-induced copolymerizations encompass a wider range of donor/acceptor combinations.

Table 2. Polymerizability of donor/acceptor pairs

D (styrenes)		Increasing acceptor strength →				
		A				
		Methyl acrylate	Itaconic anhydride	FN	N-Carboethoxy maleimide	MAn
Increasing donor strength ↓	p-methyl				trace	yes
	p-t-butyl		no	no	yes	yes
	p-(2-chloroethoxy)			yes		no
	p-methoxy	yes	yes		yes	no
	p-isoamyl			yes		no

(2) Strong donor/acceptor pairs are usually thermally too reactive for the investigation of their photo-induced copolymerizations. Spontaneous polymerizations occur beforeirradiation. Furthermore, if the donor/acceptor interaction is extremely strong, the excited complex readily dissociates to the radical-ions, which can, if the monomer is suitable, initiate ionic homopolymerization rather than radical copolymerization. If the donor/acceptor interaction is too weak, the CT complex cannot be excited to bring about the polymerization. Therefore, weak donor/strong acceptor monomer combinations or the reverse are appropriate for photo-induced radical copolymerization. This is clearly shown in Table 3: only the CT complex absorbs the light and is excited.

The same trend is also shown in Table 2. The photocopolymerization of the strongest monomer pairs such as α-MSt/MAn is slower than that of moderately strong monomer pairs such as St/MAn and 2-VN/FN.

Table 3. Initial Rp (%/min) of Photo-Induced Copolymerizations of D/A pairs[a]

	Donor Monomer	Increasing Acceptor Strength →			
		Acceptor Monomer			
		DEF	DEM	FN	MAn
Increasing Donor Strength ↓	VOAc	0 (−)[c]	0 (−)	0 (−)	0.017 (−)
	2-VN	0 (−)	0 (−)	0.21 (+)	0.14 (+)
	St	0 (−)	0 (−)	0.045 (+)	0.91 (+)
	α-MSt	0.020 (−)	0.014 (−)	0.038 (+)	0.14 (+)

[a] [D] = [A] = 1.0 M, in chloroform at 35 °C;
[b] No photo-copolymerization was observed;
[c] Results of control experiments to detect thermal copolymerizations, conditions: [D] = [A] = 1.0 M, 80 °C, in chloroform, 48 h, in the dark. (+): thermal copolymerization was observed, (−): no thermal copolymerization.

(3) Compared to excitation of the CT complex, excitation of monomer is a more effective pathway for polymerization or cycloaddition.

The range of polymerizable combinations was found to be even wider if a third monomer such as AN was added. For some weaker-D/weaker-A pairs such as VN/DEF, terpolymerization actually occurred in AN as solvent. So it is proposed that in these systems initiating radicals were formed which then were trapped by the very active monomer AN, resulting in chain propagation.

2.2 Further Discussion of Exciplexes

2.2.1 Emitting and non-Emitting Exciplexes

Exciplexes may be observed by means of fluorescence spectrophotometry. Many exciplexes show a new broad, non-structural emission at longer wavelength, while in other cases no such exciplex emission was observed except for the fluorescence quenching of the fluorescent partner. The non-fluorescent exciplex is attributed to some faster deactivation processes, such as dissociation into radical-ions (especially in polar solvents), intersystem crossing or chemical reaction. For example, in the 2-VN/FN [26] system the exciplex emission was observed, while in St/FN or 2-VN/MAn only the fluorescence quenching of St and VN occurred by mixing the partners.

2.2.2 Polarity of Exciplex

Exciplexes may possess different polarities, that is, with different charge-transfer character, which will affect their behavior. Three types of exciplexes can be distinguished: non-polar, polar and exciplexes with complete charge-transfer (intimate ion pairs). Cycloadditions occur more efficiently from relatively non-polar exciplexes than from highly polar exciplexes [39]. With increasing solvent polarity the lifetime of polar exciplexes decreases owing to the formation of solvated radical ions and the quantum yield of cycloaddition is also reduced. With exciplexes of low polarity, however, reasonably high quantum yield of cycloaddition can be achieved even in polar solvents. In the photoreaction of anthracene with dienes less polar exciplexes favor the Woodward-Hoffmann-allowed $\pi 4s + \pi 4s$ cycloaddition, whereas polar exciplexes yield the $\pi 4s + \pi 2s$ adducts [35]. Obviously, the polarity of the exciplex is an important factor to be considered in the initiation of copolymerizations. Intimate ion pairs usually show different behavior: they undergo heterolytic fission to form radical pairs or they decay via the formation of the triplet state.

2.2.3 Triplet and Singlet Exciplexes

Exciplexes are usually considered to be singlet, but triplet exciplexes have also been found. Many photoreactions proceed by way of the triplet state. In these cases, the triplet exciplex may be involved. A triplet state can be produced by three pathways: 1) intersystem crossing of the singlet state of the exciplex or one of the partners; 2) recombination of the radical-ion pairs or back electron-transfer; 3) energy transfer by triplet sensitizers.

There has been a tendency in the literature to infer that triplet excited state complexes are as important as singlet exciplexes [40–43]. The evidence, unfortunately, has been much more tenuous to obtain. Direct observation is difficult since triplet exciplex phosphorescence is extremely rare [44]. Corey [45], utilizing substituent effects on relative reactivity, years ago inferred the intermediacy of a triplet complex in the addition of alkenes to cyclohexenone. Other evidence usually comes from the regiochemistry. In some addition reactions, the product structures are incompatible with the most stable diradical and this suggests that there is a precursor to the diradical. For example, in the reaction of acetone with vinyl ether, a simple diradical pathway should give almost solely the 3-alkoxyoxetane (**4**), if the stability of the possible diradical intermediates is dominant in determining the reaction pathway; diradical (**2**) should be favored over diradical (**1**). In fact, the reaction shows little regioselectivity suggesting that a triplet exciplex preceeds diradical formation [46].

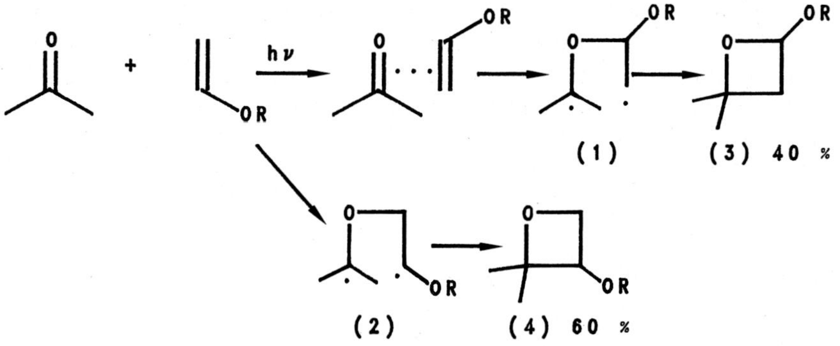

(11)

2.2.4 Triplexes

Quenching of the exciplex fluorescence by a third molecule may lead to a termolecular complex called triplex. It was first observed by Beens and Weller [47] and has since been shown to be a general phenomenon [48–53]. Each exciplex shows a marked preference for quenching by donor (D_Q) or acceptor (A_Q). The species approximately represented by $D^{\delta+} \ldots A^{\delta+}$ should be quenched from the D side by D_Q, i.e., $D_Q^{\delta+} \ldots D^{\delta+} \ldots A^{\delta-}$, or from the A side by A_Q, i.e., $D^{\delta+} \ldots A^{\delta-} \ldots F\, A_Q^{\delta-}$. Caldwell's results showed a quenching preference when the interaction occurred with the exciplex component of lower singlet energy. Thus the exciplex of 9-cyanophenanthrene (9-(NP) and p-(isobutenyl) anisole (p-BA) clearly prefers A_Q, while the exciplex of phenanthrene and FN prefers D_Q.

A triplex has been proposed as an intermediate in some photoreactions. Yang and co-workers [54] found that addition of 1,3-dienes to anthracene excimers leads to different products than does its addition to monomeric excited arenes. Lewis and co-workers [55, 56] found that stilbene excimers can be intercepted by dimethyl fumarate to give an oxetane through a presumed triplex. More recently, Schuster and co-workers [57, 58] studied the "Triplex Diels-Alder" reaction of 1,3-dienes with enol, alkene, and acetylenic dienophiles. Take 9,10-dicyanonaphthalene (DCN)/indene (IN)/1,3-cyclohexadiene (CHD) system as an example:

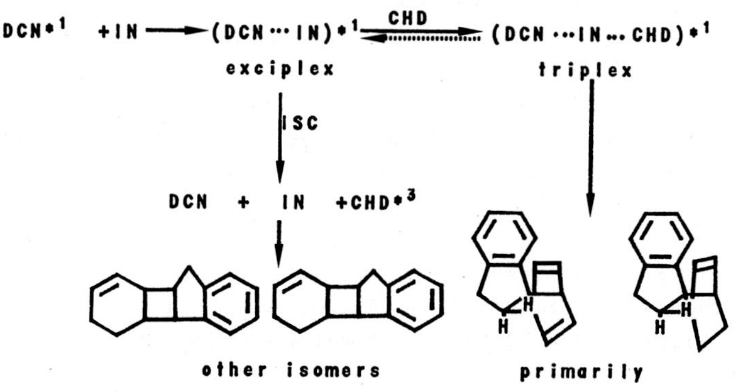

(12)

The excited singlet DCN combines with IN to form an exciplex. It can become a triplet exciplex through intersystem crossing and eventually generate the triplex CHD (the compound with lowest triplet energy). Alternatively, it can be captured by a second donor (CHD) to form a triplex, which was suggested to be the precursor of the Diels-Adler adducts. The spectral measurements indicate a high degree of charge-transfer from IN to DCN in the exciplex. Thus IN is converted from a rather electron-rich compound to a decidedly electron-poor one by complexation with DCN. This effect is identical with the catalysis effect of Lewis acids to some Diels-Alder reactions.

2.3 Internal Chemistry of Exciplexes

Exciplexes can undergo different chemical reactions. For the donor olefin/acceptor olefin pairs, there are four possibilities from the point of view of bond-formation:
(1) formation of two C—C bonds: concerted cycloaddition;
(2) formation of one C—C bond: tetramethylene formation;
(3) No C—C bond formation: single electron transfer (SET) to ion-radicals;
(4) Loss of one C—H bond: proton or silyl transfer, the cleavage of original C—H bond.

As to reaction with external reagents, the work of Schuster cited above is perhaps the most clearly defined example.

2.3.1 Formation of Two C—C Bonds: Concerted Cycloaddition

Excimers and exciplexes had long been postulated as intermediates in photocycloadditions. Nevertheless, only after Caldwell's outstanding work [59–61] was it verified that the emitting exciplex is indeed the precursor to the singlet cycloaddition and related reactions, rather than an energy-wasting byproduct. For example, in benzene solution, exciplex fluorescence from (10-cyanophenanthrene) (**5a**) and (β-methylstyrene) (**6a**) is observed with λ max = 435 nm. This fluorescence is quenched by adding dimethyl acetylenedicarboxylate, with a Stern-Volmer slope $K_{sv} = 64\ M^{-1}$. The cycloaddition reaction is quenched with $Ks_v = 62\ M^{-1}$. This result implies that the emitting exciplex is an intermediate in the cycloaddition reaction or is in equilibrium with such an intermediate [60, 62]. The former explanation is in reasonable agreement with theory [63, 64], and with experimental results on other singlet additions.

Using the temperature effect, it was proven that non-emitting exciplex intermediates were also involved in the cycloaddition. For example, in the reaction of naphthalene with diphenylacetylene, with increasing temperature both the quantum yields of photoaddition and the quenching of the naphthalene fluorescence by acetylene decrease by the same magnitude [65].

An exciplex can exert a favorable effect on cycloaddition reactions in two ways. First, the prior association favors the bimolecular process. Second, exciplex formation will bias the system toward bond-formation of the donor-acceptor pair, and away from competing chemical processes, such as dimerization of the donor.

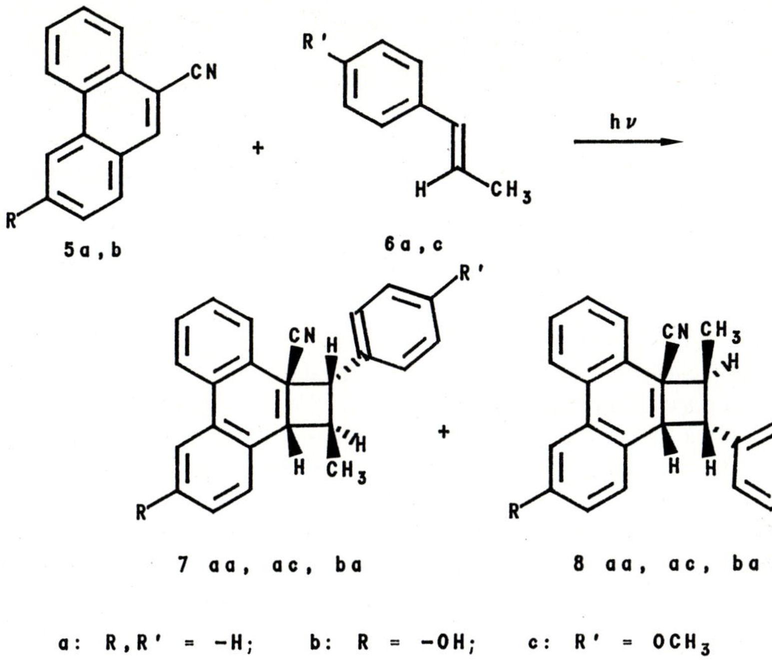

[Scheme 13: Photocycloaddition of 5a,b with 6a,c giving 7 aa, ac, ba and 8 aa, ac, ba]

a: R, R' = −H;   b: R = −OH;   c: R' = OCH₃   (13)

For a further discussion of possible Woodward-Hoffmann allowed photocycloadditions, see Sect. 4.2.

2.3.2 Formation of One C−C Bond: Tetramethylene Intermediates

In recent years, more and more photochemists consider that many photocycloaddition reactions take place stepwise, that is, one bond is first formed. The formed tetramethylene intermediate may close to a cyclobutane or initiate the polymerization. We will leave this discussion to Sect. 4.

2.3.3 No C−C Bond Formation: Ion-Radical Formation

With increased polarity of the medium, an exciplex with moderate to strong charge-transfer character tends to dissociate into radical ions. Radical ions have been extensively observed and studied by virtue of flash photolysis. Several follow-up reactions may occur. Reverse electron transfer between ion pairs, either in the cage or through recombination, leading to the ground state, is an unfortunate energy-wasting step responsible for the low quantum yields in many electron-transfer reactions. Recombination of radical ion pairs can also lead to the triplet state of one of the partners.

Among the first examples of photo-induced reactions through radical-ion intermediates was the dimerization of VCZ, discovered by Ellinger [66], and later thoroughly investigated by Ledwith [67, 68].

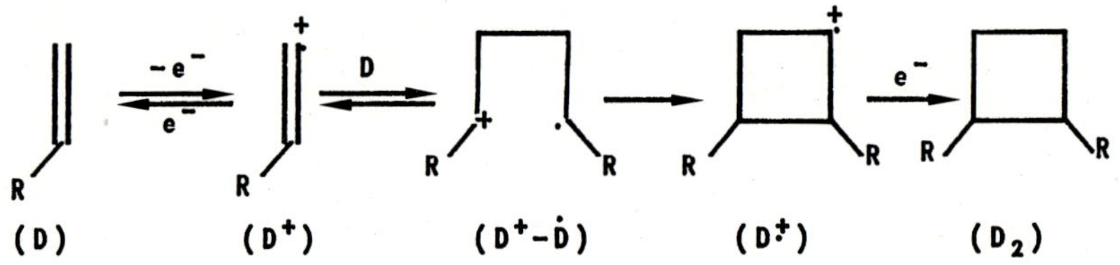

$$\text{(D)} \xrightleftharpoons[e^-]{-e^-} \text{(D}^+\text{)} \xrightleftharpoons{D} \text{(D}^+\text{-}\dot{\text{D}}\text{)} \longrightarrow \text{(D}^{\ddagger}\text{)} \xrightarrow{e^-} \text{(D}_2\text{)} \tag{14}$$

a: R = Carbazole 62% [66, 67, 68]
b: R = OPh 30% [69–70]
c: R = Aryl 10–30% [71, 72]

Here a secondary electron transfer between the radical cation and a neutral donor molecule produces a 1,4-cation radical. The acyclic 1,4-cation radical is in equilibrium with a cyclobutane radical cation. Other dimerizations have been described by Farid et al. [73–75], Arnold et al. [76], Pac et al. [77, 78], and others [79–81].

In the last step, the electron transfer to $D_2^{\cdot+}$ can be from a molecule, D. This constitutes a chain reaction resulting in high quantum yield, occasionally exceeding one:

$$D^{\cdot+} + D \rightleftharpoons D_2^{\cdot+}$$
$$D_2^{\cdot+} + D \rightleftharpoons D_2 + D^{\cdot+} \tag{15}$$

Noteworthy is that in this process a bond is formed only between two donor molecules. This makes it easy to distinguish an ion-radical process, formation of the homodimer product from a possible zwitterion intermediate which would produce a heterodimer. A few examples of additions of the radical cation to a different donor olefin are also known [74].

Occasionally in the intimate radical ion pair, reverse electron transfer from A to a 1,4-cation radical may form a diradical.

$$A^1 + D \longrightarrow A^{\cdot-}/D^{\cdot+} \xrightarrow{D} A^{\cdot-}/D\text{-}D^{\cdot+} \longrightarrow A + \dot{D}\text{-}\dot{D} \longrightarrow D_2 \tag{16}$$

1,4-Cation radicals have both radical and cationic nature. With the former, they can react with oxygen, leading to the oxidation products:

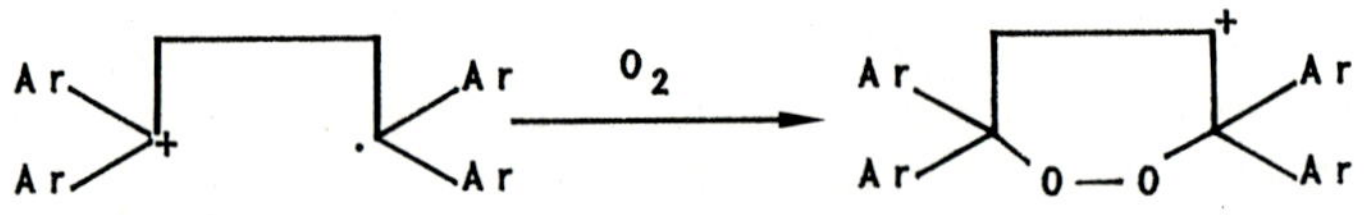

$$\text{Ar}_2\text{C}^+\text{-}\dot{\text{C}}\text{Ar}_2 \xrightarrow{O_2} \text{Ar}_2\text{C}\text{-}\text{CAr}_2\text{(O-O)}^+ \tag{17}$$

and with the latter, they can react with methanol, leading to the methanolated products as follows:

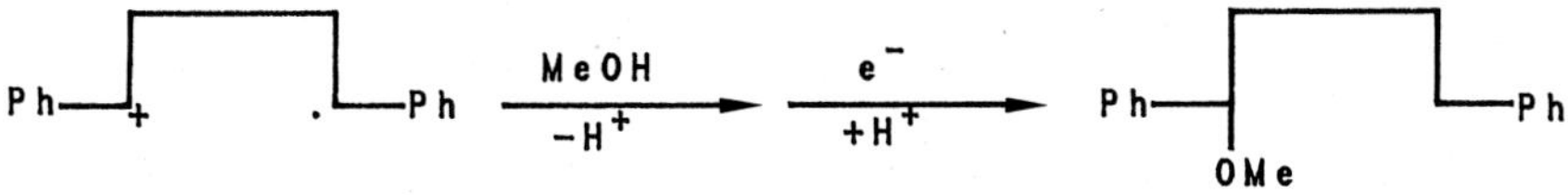

$$Ph-\!\!\!\sqcap\!\!\!\sqcup\!\!\!-Ph \xrightarrow[-H^+]{MeOH} \xrightarrow[+H^+]{e^-} Ph-\!\!\!\sqcap\!\!\!\underset{OMe}{\sqcup}\!\!\!-Ph \quad (18)$$

The tetramethylethylene/2-cyanonaphthalene system undergoes cycloaddition in benzene to produce stereospecific product as follows:

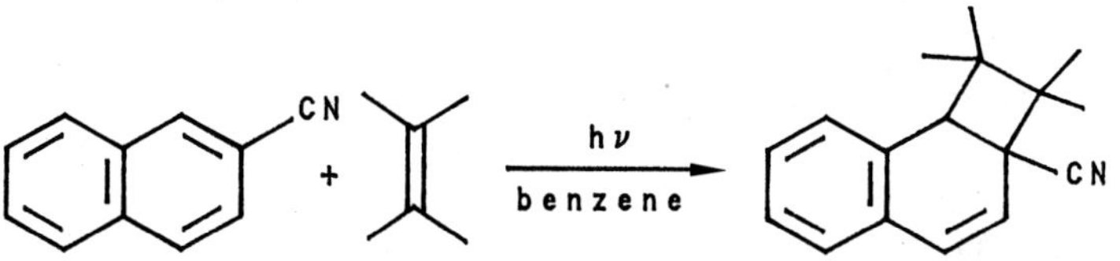

$$\text{naphthalene-CN} + \text{tetramethylethylene} \xrightarrow[benzene]{h\nu} \text{cycloadduct-CN} \quad (19)$$

However, if it is irradiated in methanol, no cycloaddition occurs, but a linear adduct is formed by the radical ion mechanism shown in Eq. (20) [83–83].

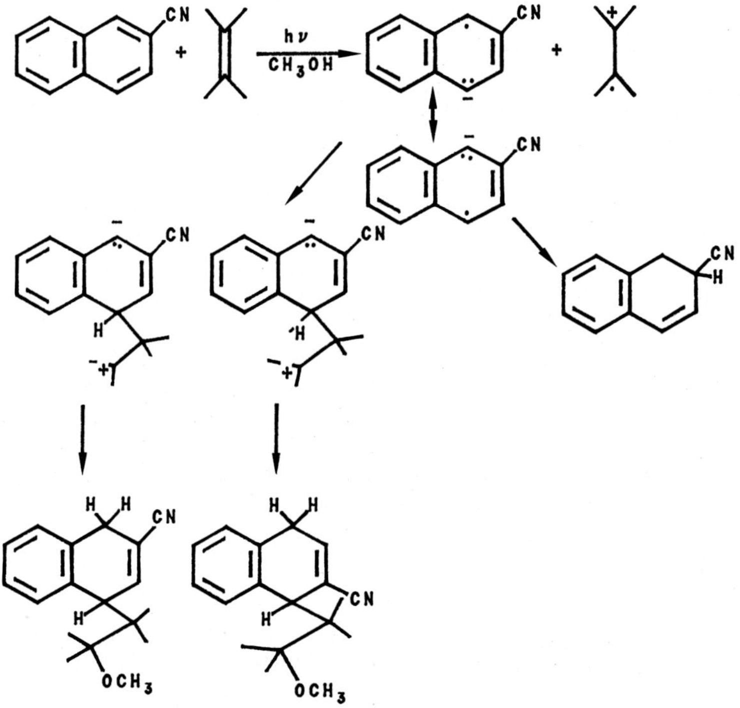

(20)

Shirota [5] has postulated a scheme involving such "dimer cationic radicals" as the intermediates for cyclodimerization and polymerization of VCZ. The ratio of the competing reactions depends mainly on the solvent basicity. In strong basic solvents only radical polymerization occurs, while it is accompanied by cyclodimerization of VCZ in moderate basic solvents and by cationic homopolymerization of VCZ in non-basic solvents.

These homodimerizations of the donor olefins are well accounted for by the ion-radical chain mechanism, and provide powerful evidence for the occurrence of single electron transfer. Were it not for these cyclodimers, ionic homopolymerizations of D or A could be interpreted as initiation by zwitterionic tetramethylenes formed from D and A.

2.3.4 Loss of One Bond: Proton Transfer Reaction

The proton transfer reaction, in which no $C-C$ bond but instead a $C-H$ bond is formed, is known to be an important follow-up reaction of the exciplex when the donor or the acceptor possesses active hydrogen atoms. It proceeds in the intimate radical ion pairs or between the radical ions. The first charge-transfer, followed by proton transfer, produces two free radicals, which may initiate polymerizations or lead to addition products. Lewis [84] has recently reviewed this reaction.

The first such reaction to be described in the literature was the substitution at the benzylic $C-H$ bonds of indene, which occurs upon irradiation of indene and acrylonitrile [85]. This reaction involves electron transfer followed by proton transfer and radical coupling:

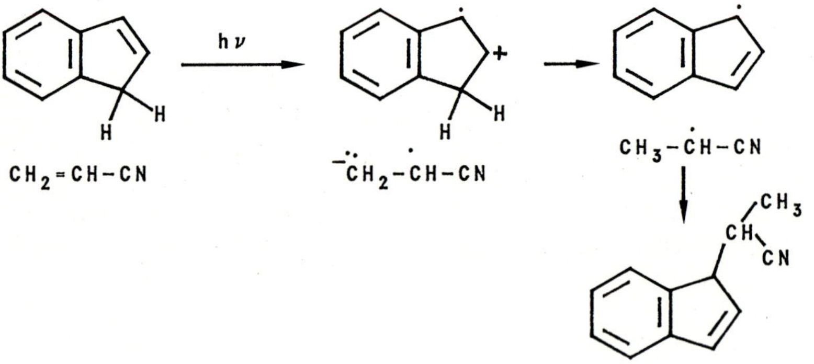

$$\text{(21)}$$

Indenes lacking benzylic hydrogens (e.g., 1,1-dimethylindene) give no substitution products upon irradiation with acrylonitrile.

3 Tetramethylenes in Spontaneous Thermal Reactions of Donor/Acceptor Olefin Pairs

Before discussing the tetramethylenes in the photoreactions of donor/acceptor monomer pairs, we would like to recall the results of thermal copolymerizations of such pairs, from which enlightment may be gained for our proposal. In all discussions we will deal only with spontaneous polymerizations, i.e., in the absence of initiators.

Many donor-acceptor monomer pairs spontaneously undergo thermal copolymerizations. Hall [86–88] studied the thermal addition and polymerization reaction of such combinations systematically and postulated a "Bond-Formation Initiation" theory. The most important points of this theory may be summarized as follows:

1) Electron donor-acceptor monomer pairs form charge-transfer complexes (CTC), which collapse to the tetramethylene intermediates through the bond-formation between the β-carbons:

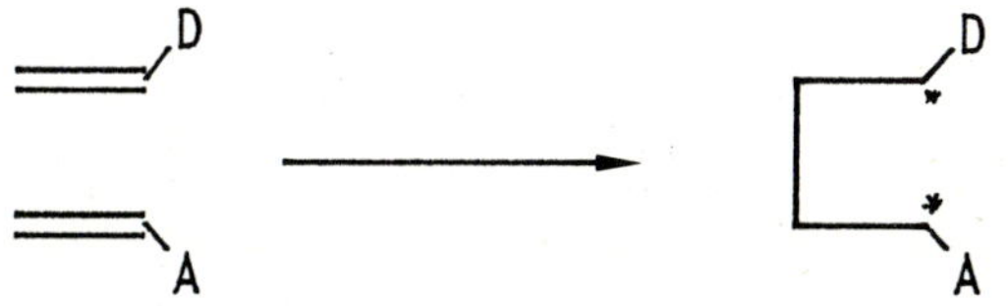

$$\begin{array}{c}\diagup\!\!\!\!\!\!=\!\!\!\diagup^D \\ \diagdown\!\!\!\!\!\!=\!\!\!\diagdown_A\end{array} \quad \longrightarrow \quad \begin{array}{c}\diagup^D_* \\ \diagdown_A^*\end{array} \qquad (22)$$

∗,∗ = +,− or ·,·

where D and A represent donor and acceptor substituents, respectively. This tetramethylene acts as the common intermediate for polymerization as well as for the formation of small molecule products, especially cyclobutanes.

2) The tetramethylene is a resonance hybrid of 1,4-diradical (∗,∗ = ·,· in Eq. (22)) and zwitterionic (∗,∗ = +,−) limiting structures. The character of the tetramethylene is determined by the nature of the terminal substituents. A very strong donor substituent at one of the terminal carbons and a very strong acceptor substituent at the other leads to zwitterionic intermediates. Otherwise, for instance, phenyl or vinyl group at the donor terminal and diester, cyano-ester or anhydride at the acceptor terminal, will favor the diradical form.

3) β-Substituents have no influence on the diradical or zwitterionic character of tetramethylene. However, their presence favors the *gauche* conformation over the *trans* (extended) form [89]:

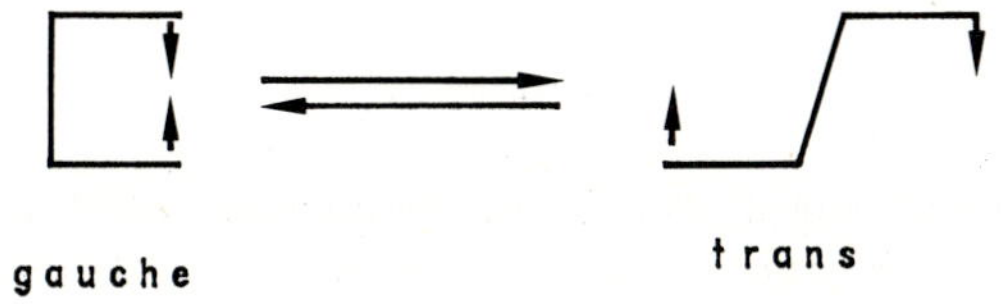

$$\text{gauche} \quad \rightleftarrows \quad \text{trans} \qquad (23)$$

and thus favors the cycloaddition. In addition, steric hindrance by the β-substituents will retard bond formation at that position.

4) The high reactivity of these tetramethylene intermediates has generally enabled them to escape direct detection and isolation. An extremely effective way of trapping them is to initiate polymerization. The nature of the polymerization will diagnose the diradical or zwitterionic character of the tetramethylene. Several investigators have trapped zwitterionic tetramethylene by using methanol, acetone or acetonitrile [87], and there is an example of a trapped diradical tetramethylene with 2,2,6,6-tetramethylpyperidine N-oxyl (TEMPO) [88], although this is much less common.

5) In cases in which a pair of reactants undergoes both cyclization and polymerization the contrasting results are controlled by the experimental conditions, relative concentrations and the mode of work-up. "Organic Chemists' reaction conditions" use lower concentration of the reactants, usually in equimolecular amounts, which favor the cycloaddition products, while "Polymer Chemists' reaction conditions" use higher concentration of the reactants, one of which is usually excess, and thus favor polymer products.

6) A zwitterionic tetramethylene initiates ionic homopolymerization, while a diradical tetramethylene initiates free radical copolymerization. As initiating species, zwitterions are likely to remain in the coiled gauche-conformation and collapse to small molecules. Diradicals, on the other hand, are easily transferred to the trans-conformation. Accordingly, diradicals are more effective initiators and more radical copolymerizations occur than ionic homopolymerizations. Addition of solvent will also influence the reaction of polar tetramethylene. A polar non-donor solvent may permit carbenium ion polymerization, while a polar donor solvent impedes it.

7) The most important competing process to the bond-formation is the complete electron transfer to form ion-radicals, which occurs where no bond formation is possible, for example, for aromatic donor-acceptor pairs. For vinyl copolymerizable pairs, the bond will form between the components to give a diradical tetramethylene. For the ionic homopolymerization system, on the other hand, it is difficult to distinguish the ion-radicals from zwitterionic tetramethylenes by the kinetic analysis. In this case, the accompanying cycloaddition reaction offers powerful evidence for the zwitterion formation, i.e., the bond-formation.

8) The effect of the donor and acceptor substituents on the reactivity and mechanism of tetramethylene in both organic and polymer reactions can be arranged as an "Organic Chemist's Periodic Table", wherein the areas of mechanistic change clearly emerge and which provides predictive capability.

Thus, Hall has proven that tetramethylene intermediates, arising from bond-formation between the β-carbons of reacting olefins, are the key to small molecule formation and the thermal copolymerization. This unifying concept of bond-forming initiation has been also extended to spontaneous addition and polymerization reaction of heteroatom acceptor molecules and 7,7,8,8-tetrasubstituted quinomethanes.

A clear case of cationic homopolymerization initiated by zwitterionic tetramethylene was thoroughly investigated in the reaction of VCZ with tetrasubstituted electrophilic olefins [88]. When "Polymer Chemists' Conditions" produced VCZ homopolymer, "Organic Chemists' Conditions" gave a cyclobutane, which could in turn initiate the polymerization of VCZ. The proposed mechanism involves formation of CT complex and then bond-formation to form the zwitterionic tetramethylene, which closes reversibly to the cyclobutane adduct, or initiates cationic homopolymerization of VCZ. The zwitterionic tetramethylene was trapped very efficiently with methanol and internal trapping was also observed.

Similarly, zwitterionic tetramethylenes as initiators of anionic polymerization were also observed. For example, methyl α-cyanoacrylate polymerizes via an anionic mechanism in the presence of n-butyl vinyl ether [90]. A Diels-Alder adduct is also formed. In another example, the reaction of isobutyl vinyl ether and nitroethylene leads to an unstable adduct [91], which is capable of initiating the anionic polymerization of excess nitroethylene, and also the cationic polymerization of added VCZ.

Diradical tetramethylene as the initiating species in the free radical copolymerization was studied in detail in the dimethyl cyanofumarate/p-methoxystyrene pair [87]. As the monomer concentrations decreases to very low values, a dihydropyran cycloadduct began to appear at the expense of copolymer. The yield of the small molecule product was favored by high dilution, implying the competitive formation of polymer and dihydropyran from a common intermediate-tetramethylene. At low monomer concentration, the chance of a monomer reacting with the diradical tetramethylene is smaller, the radicals stay in the "cage" and react with each other to form the cycloadduct. At higher concentrations, the tetramethylene diradical is trapped by monomer molecules so the polymerization takes place. The diradical tetramethylene was trapped by using TEMPO (2,2,6,6-tetramethylpyperidine N-oxyl). A 1:1:1 adduct could be isolated, thereby proving that the tetramethylene is really present in the reaction mixture. Another evidence for tetramethylene diradical was relationship between polymer molecular weight and polymerization time. If initiation and propagation at diradical were actually occurring, one would expect an increase in polymer molecular weight with the increase of the conversion. This is actually observed in the studied system, and strongly supported the diradical initiation mechanism.

We shall utilize the above insights about the behavior of tetramethylene diradicals in our discussion of photoreactions of donor/acceptor olefins.

4 Tetramethylene Diradicals as Key Intermediates in Photoreactions of Donor/Acceptor Olefins

4.1 Tetramethylene Diradicals in Photochemistry

The tetramethylene diradical or 1,4-diradical has long been an attractive subject in photochemistry. During the past twenty years it has become possible to obtain direct experimental information bearing on these species, and with the accumulation of these data we can now begin to understand the factors which influence diradical behavior.

The Norrish Type II reaction is well known to produce a wide variety of diradicals, including 1,4-diradicals; such as in the case of the following scheme:

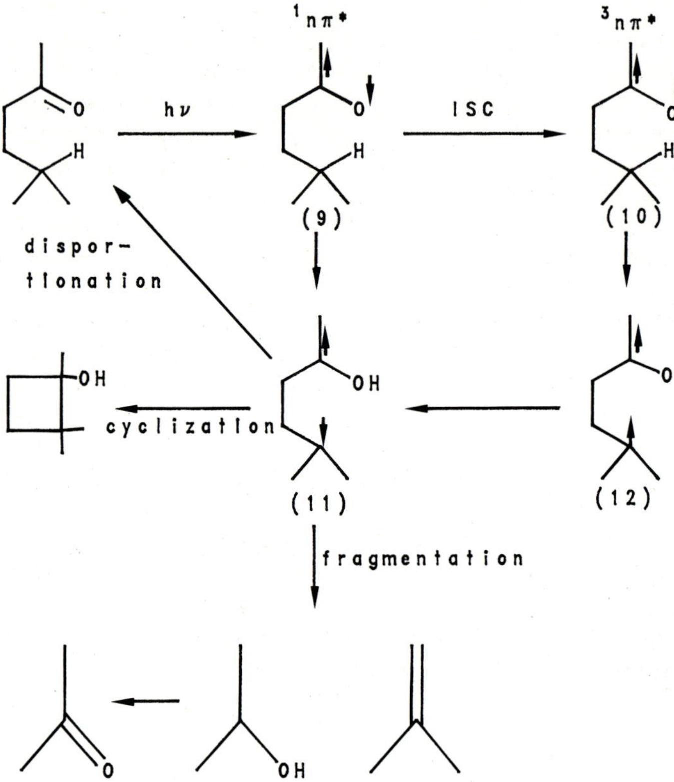

(24)

The singlet n, π* carbonyl excited states, especially those of aryl carbonyls, undergo rapid ISC to the triplet manifold. Then the diradical is formed through the intramolecular abstraction of a tertiary hydrogen atom by the excited carbonyl species [92]. While this abstraction may occur from either the singlet or triplet n, π* carbonyl excited state (**9** or **10**, respectively) to form either the singlet or triplet Type II diradical (**11** or **12**, respectively), the triplet diradical has been demonstrated to have much longer lifetime which seems to be governed by the rate of ISC to the singlet diradical **11**. Once formed, **11** rapidly undergoes one of the following transformations: 1) disproportionation with reverse hydrogen transfer and return to the starting carbonyl compound; 2) cyclization to cyclobutanol; 3) fragmentation to form an olefin and a carbonyl compound.

Norrish Type II diradicals can be intercepted by olefins, albeit in very poor yields. Polymerization of neat MMA can be initiated photochemically with alkyl phenones [93].

The Paterno-Büchi reaction between excited carbonyl groups and olefins leads to the formation of four-membered cyclic ethers (oxetanes). This reaction has been reviewed by Arnold [94], Jones [95], Wilson [96]. The scope of the Paterno-Büchi reaction is impressive. It might be initiated by either singlet or triplet, n, π* or, less frequently π, π* excited carbonyls, and both electron-rich as well as extremely electron-poor olefins participate. It is now widely accepted that the initial intermediate in the oxetane formation is an exciplex between the excited carbonyl triplet and the olefin. The exciplex is then thought to decay to the triplet 1,4-diradical which, upon ISC to the singlet diradical, may either fragment back to the starting components or cyclize to oxetanes.

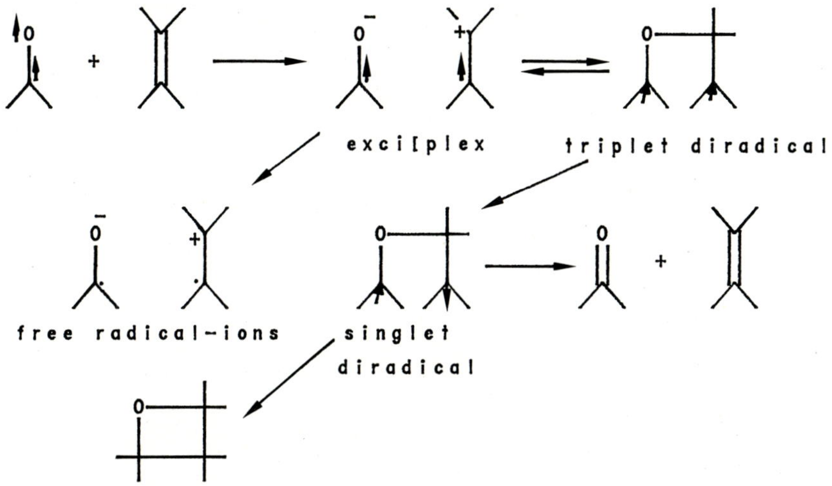

(25)

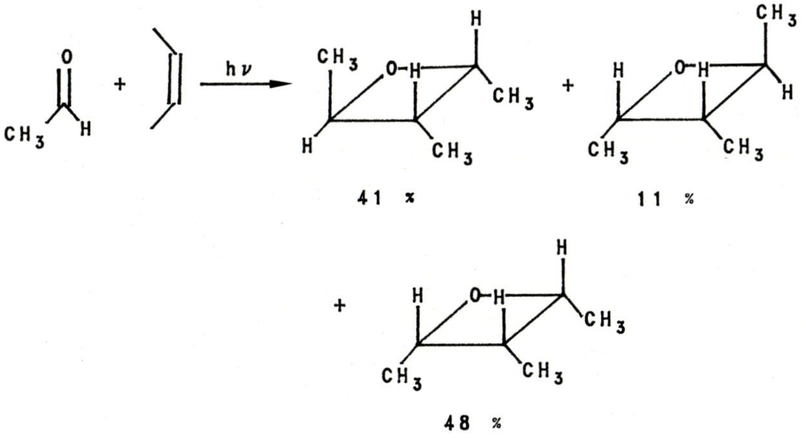

(26)

Yang et al. [97] examined the stereochemistry of the photocycloaddition of acetaldehyde and *cis*- and *trans*-2-butene which they suggested proceeds via the n, π* state. A mechanism involving singlet 1,4-diradical was proposed to explain the 4% loss of stereochemistry in the addition to *trans-* and the 11% loss of stereochemistry in addition to *cis*-butene.

The oxetane formation is controlled by the easy formation of the exciplex which may subsequently collapse to give the diradical intermediate. The proposed sequence of steps is as follows:

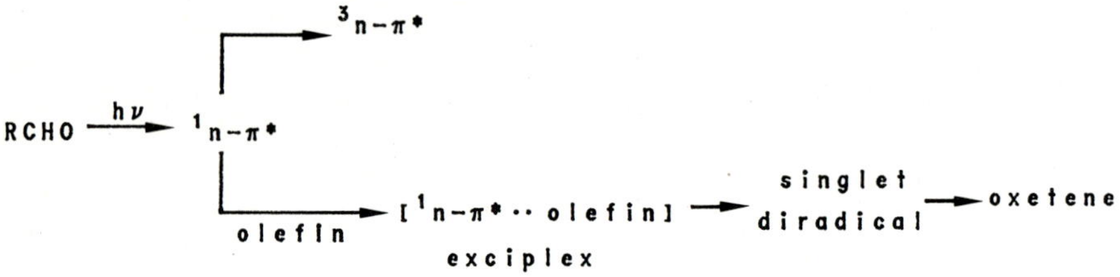

$$RCHO \xrightarrow{h\nu} {}^1n-\pi^* \begin{array}{c} \longrightarrow {}^3n-\pi^* \\ \downarrow \text{olefin} \end{array} [{}^1n-\pi^* \cdots \text{olefin}] \longrightarrow \begin{array}{c} \text{singlet} \\ \text{diradical} \end{array} \longrightarrow \text{oxetene}$$
$$\text{exciplex}$$

(27)

The photolysis of cyclic azoalkanes provides another useful source of diradical generation [98–100]. One example is the 1,4-cyclohexadiyl diradical [101]:

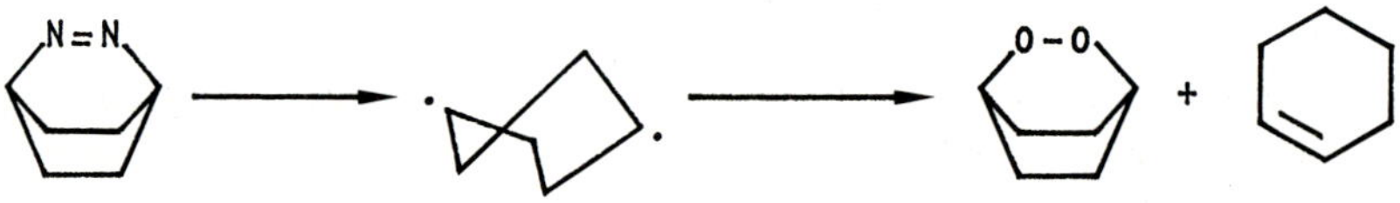

(28)

At 10 atm of oxygen about 4% of the triplet 1,4-diradical generated could be trapped.

4.2 Tetramethylene Diradicals as the Intermediates in Cycloadditions of Donor/Acceptor Olefins

Cycloaddition reactions of α,β-unsaturated carbonyl compounds to olefins have been studied in detail. Cyclic enones, undergo rapid and efficient intersystem crossing thereby providing easy access to the triplet state through direct excitation. The following cycloaddition reaction is proposed to involve a diradical intermediate formed directly or from an exciplex [102, 103] (see (29)).

If the reaction proceeds directly to a diradical intermediate, the regioselectivity with an unsymmetrical alkene will be determined from the relative stability of the possible diradical intermediates. Diradical **13** will be less stable **14**. If, on the other hand, the reaction proceeds from the exciplex, the orientation of alkene in the

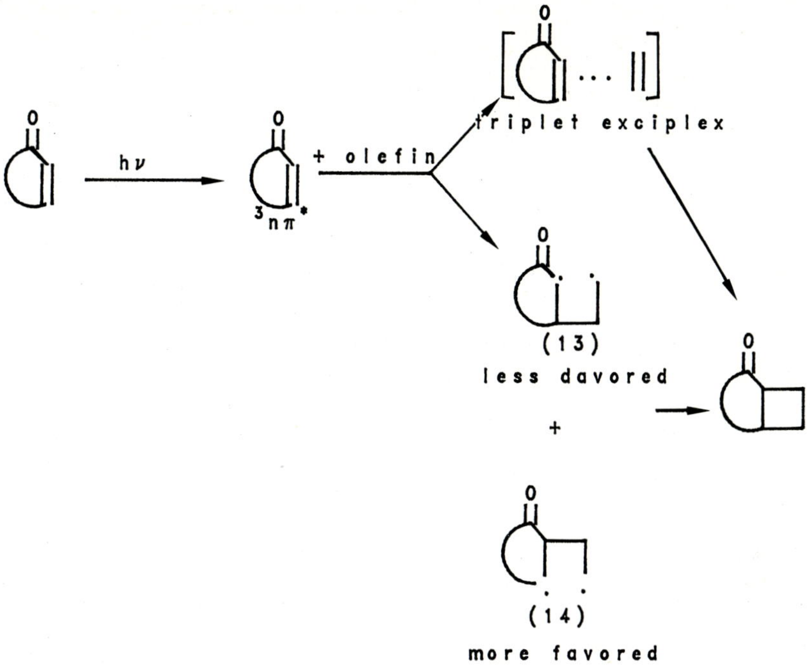

The Role of Tetramethylene Diradicals

$$\text{(29)}$$

exciplex will dictate the regioselectivity of the adduct. For example, the reaction of cyclohexenone with dimethoxyethene produces head-tail cyclobutanes (mainly *anti*-adduct dictated by steric factors):

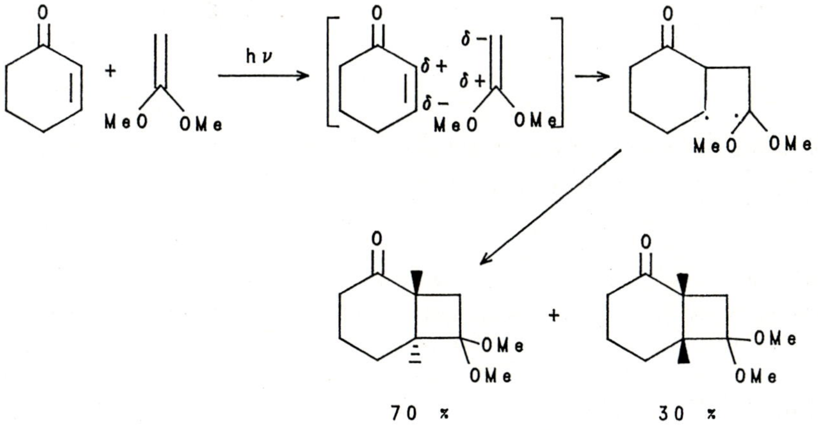

$$\text{(30)}$$

With an electron-acceptor olefin, methyl acrylate, the head-to-head product will be obtained (see (31)).

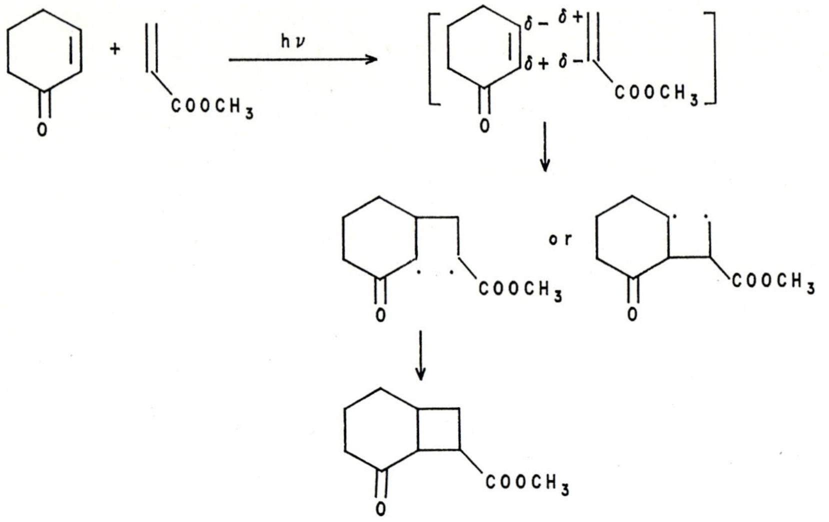

(31)

The photoreaction of 1,4-naphthoquinones with 1,1-diarylethenes [104] is as follows:

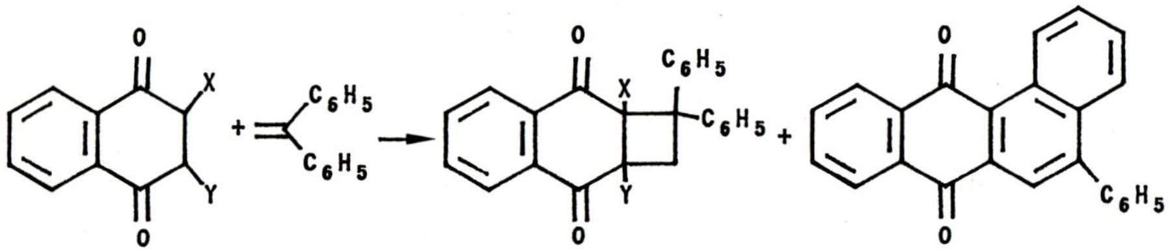

(32)

The proposed intermediacy of diradicals or zwitterions is supported by CIDNP studies.

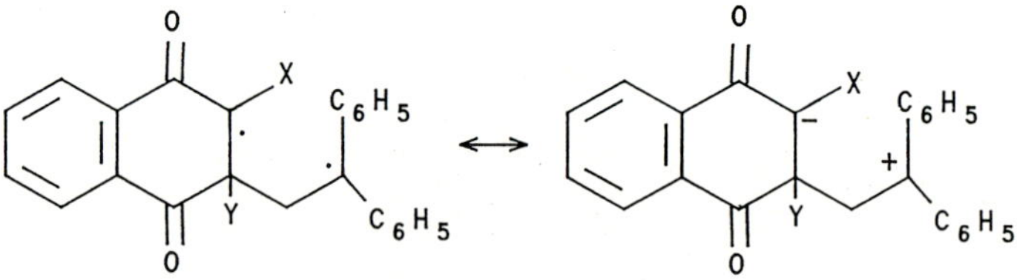

(33)

The tetramethylene diradical was also proposed as the intermediate for dimerization of olefins and [2 + 2] mixed additions between olefins. Hospool [105] suggested that if a triplet mechanism was operative, the produced diradical intermediate would undergo free rotation before the closure of the final bond and many products would be obtained. However, if a singlet state mechanism is operative, then a stereospecific addition reaction is likely to occur. Such a concerted process will fall within the compass of the Woodward-Hoffmann rules.

Lewis investigated the photodimerization of *cis*- and *trans*-anethole[1-(*p*-methoxyphenyl)propane, *c*-A or *t*-A] in the singlet state. Whereas the reaction of singlet *t*-A with another ground state *t*-A or *c*-A occurs with retention of stereochemistry, the dimerization of *c*-A occurs with retention at one double bond and inversion at the other. This is proposed to proceed via sequential formation of a singlet exciplex with optimized π-orbital overlap and a singlet 1,4-diradical intermediate [106]:

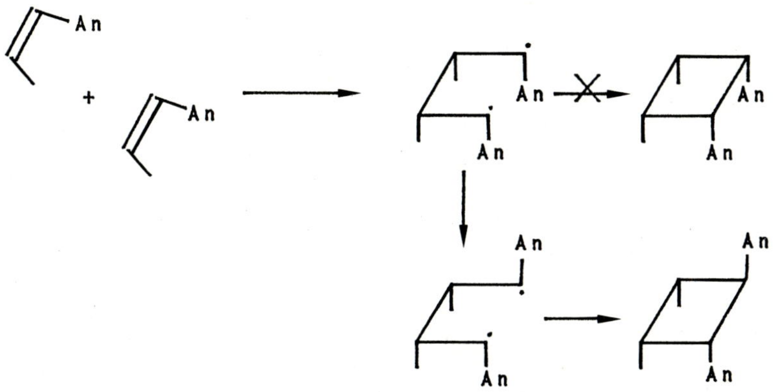

$$(34)$$

For a proposed cycloaddition mechanism, Caldwell [107] postulated two possibilities: collapse of the exciplex to a 1,4-diradical tetramethylene intermediate or direct reaction to cycloadducts:

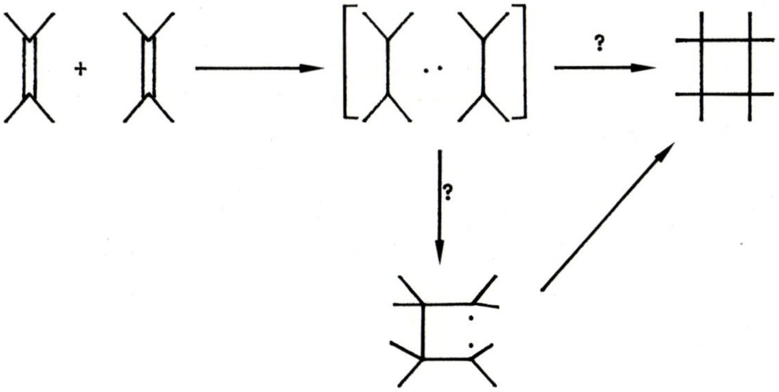

$$(35)$$

Turro [108] proposed a similar mechanism to explain the photodimerization of 2-butene: either concerted addition of $S_1(\pi\pi^*)$ to 2-butene, or formation of a singlet exciplex by interaction of $S_1(\pi\pi^*)$ and 2-butene, followed by direct collapse to cycloadducts or by collapse to a singlet 1,4-diradical which collapses to cycloadduct faster than it loses memory of its initial stereochemistry.

Lewis examined two systems, trans-α-phenylcinnamonitrile with a 2,4-hexadiene [109] and 9-cyanoanthracene with a butene [39], in an attempt to find evidence to support a proposed 1,4-zwitterionic tetramethylene intermediate. His results, including the lack of a pronounced solvent effect when the polarity was varied, failed to support such an intermediate. Curiously, he did not propose a 1,4-diradical intermediate which would not be susceptible to solvent stabilization.

In the study of photochemical cycloadditions of singlet or triplet (sensitized) diphenylvinylene carbonate with vinyl ethers [110], however, Lewis clearly proposed cycloaddition either directly from the exciplex, or via collapse of the exciplex to a diradical tetramethylene intermediate.

In the addition of dichlorovinylene carbonate to isobutylene at low temperature, sensitized by acetone, the two-step nature has been illustrated [105]. Thus, at $-20\,°C$ a normal $\pi2 + \pi2$ adduct is formed. At lower temperature $-70\,°C$, a hydrogen transfer reaction competes with the cycloaddition to give the ene adduct:

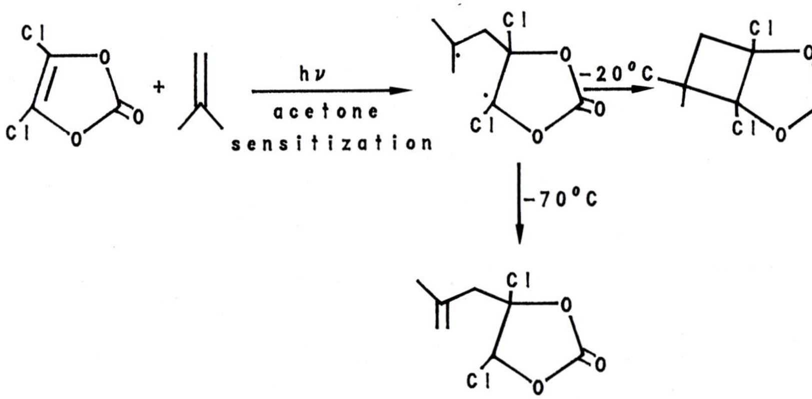

(36)

Kochi et al. [111–113] have utilized time-resolved laser spectroscopy to examine the excited complexes. For the indene (IN)/TCNE system, excitation of the CT complex (532 nm) afforded not only absorption bands assigned to TCNE anion radical and IN cation radical, but also an absorption band due to an unidentified intermediate. Kochi proposed a 1,4-diradical or 1,4-zwitterionic intermediate, and postulated bond-formation as follows [111]:

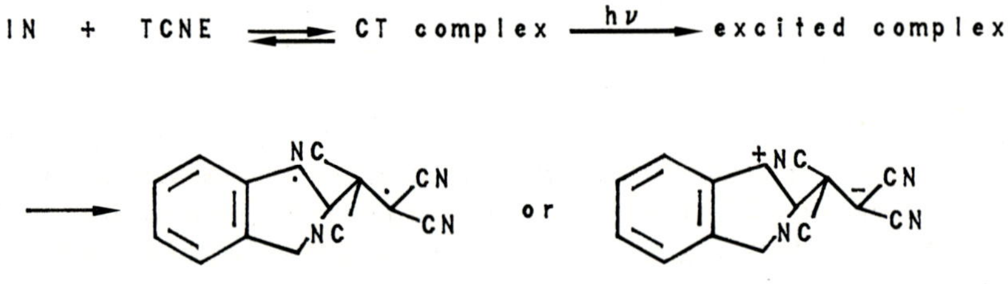

$$IN + TCNE \rightleftharpoons CT\ complex \xrightarrow{h\nu} excited\ complex$$

(37)

Mataga et al. [114] also proposed a diradical intermediate based on picosecond laser spectroscopic data. Excitation of 1,3-hexadienes/9-cyanoanthracene showed an unidentified absorption band in addition to the respective ion radicals, which might be attributed to 1,6-diradical intermediates.

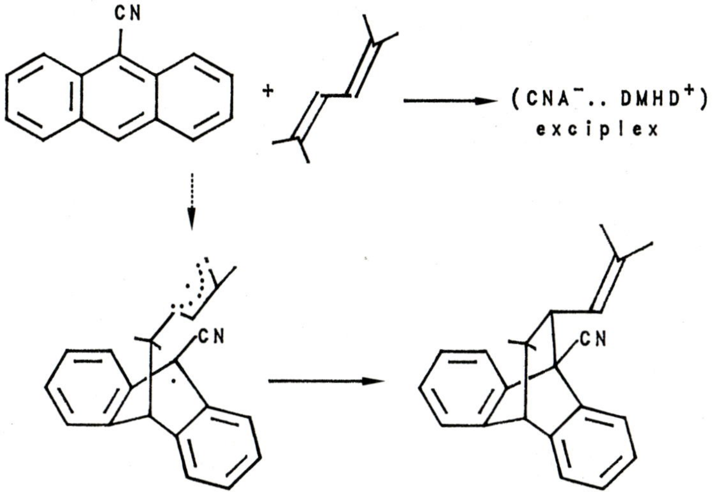

(38)

Transient absorption spectra of singlet diradical in addition to exciplex have been observed in the course of the photocycloaddition reaction of 9-cyanoanthracene (CNA) with 2,5-dimethyl-2,4-hexadiene (DMHD), while no intermediate was detected in CNA-1,3-cyclohexadiene (CHD). The cycloadduct with CHD shows an efficient adiabatic photodissociation into ^{1}CNA* and CHD immediately after excitation.

Although the mechanism of the photo[2 + 2]cycloaddition is still under investigation, one can deduce several important working hypotheses from these studies. Wender [115] proposed a scheme (see (39)).

Direct excitation of an olefin (1) generally produces a singlet excited state which can return to the ground state (−1), intersystem cross to a triplet state (2), or react with a ground state alkane to produce an exciplex (3) or a cycloadduct by a concerted path (4). The exciplex can proceed to product, ground state olefins or a diradical intermediate, which has a similar fate. When the triplet state is populated through intersystem crossing (1, 2) or through triplet sensitization (15), cycloaddition is possible via an exciplex (9), by a diradical derived from this exciplex (11) or directly from interaction between triplet and ground state alkenes (not shown). The mechanism for a given cycloaddition will then be a function of the specific substrates and reaction conditions employed. It follows from this analysis that the majority of intermolecular cycloaddition reactions proceed via a triplet excited state. Three factors contribute to this situation. The singlet excited

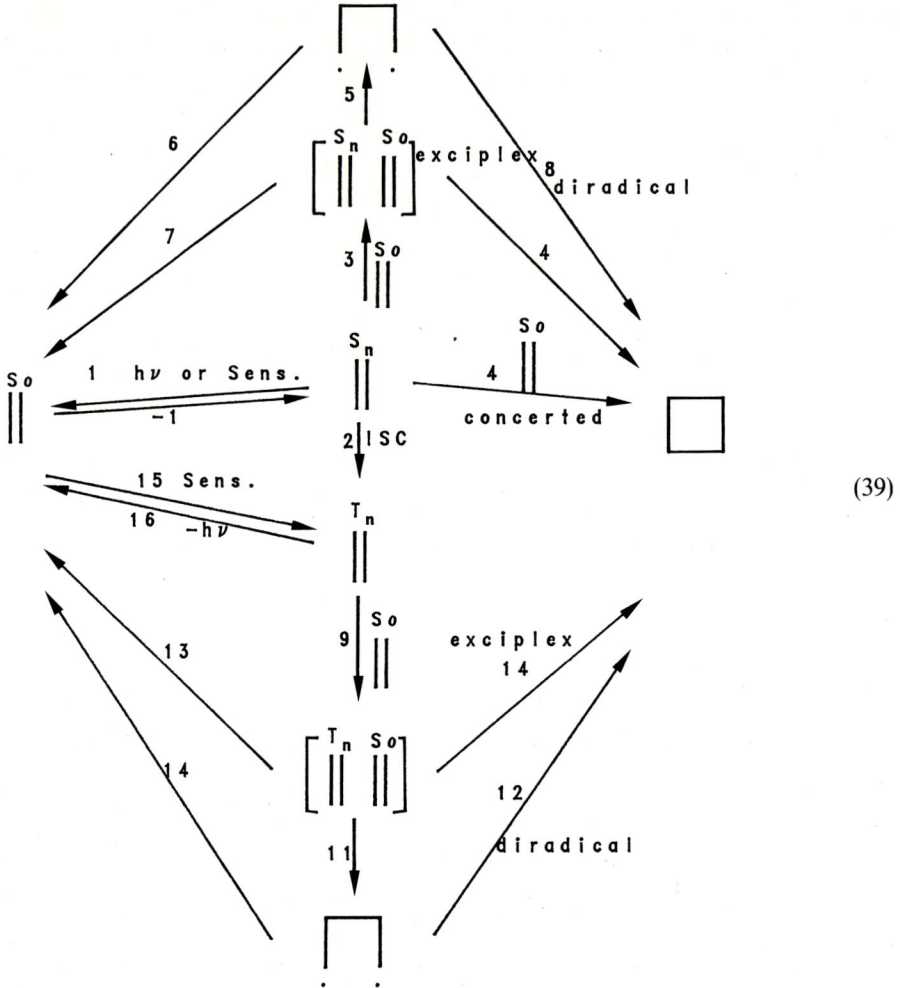

(39)

state is short-lived compared to the triplet state. For many systems, intersystem crossing to the triplet state is fast and efficient. Finally, sensitization can be used to populate the triplet state directly.

In our simplified scheme, there are four possibilites to form cycloadduct from donor olefins and acceptor olefins: (1) concerted reaction of the partners, (2) the partners form an exciplex, which collapses to cyclobutanes directly, (3) the exciplex collapses to tetramethylene diradical, which then close to cyclobutanes, and (4) direct formation of diradical from donor/acceptor pair and then cyclization.

The classical theory suggests the $[\pi_s^2 + \pi_s^2]$ cyclization reaction of a singlet olefin and a ground-state olefin is a prototypical symmetry-allowed process [116]. The concerted collapse of the olefin pairs or their exciplexes produces cyclobutanes. The evidence comes from the stereochemistry of the cycloadducts. It has been

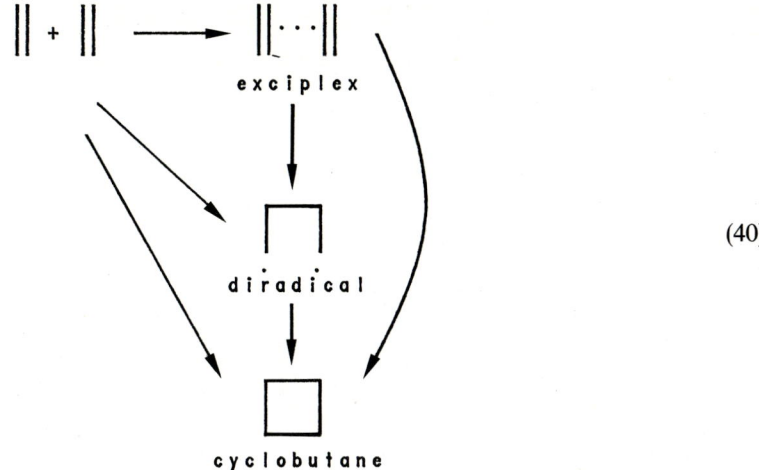

(40)

found that the major products are those having the endo structure, despite the steric hindrance. This structure reflects and maintains the maximum π-orbital overlap structure of the exciplex. The stereochemistry of the olefins (*cis* or *trans*) is retained in the adducts [117]. Numerous photocycloadditions, for instance singlet *trans*-stilbene with *cis*- or *trans*-2-butene [118], or with dimethyl maleate or fumarate [119], occur with complete retention of olefin stereochemistry. These results strongly suggest that the two bonds are formed simultaneously without any bond rotation, i.e., tetramethylene is not involved.

For the olefins with aromatic substituents, evidence for the exciplex as the intermediate has been obtained. However, for aliphatic olefins, no exciplex could be observed, so (3) is a more reasonable route.

In contrast, many photochemists have proposed 1,4-diradical tetramethylene intermediates, so (1) and (2) are also very important possible routes.

If the cycloaddition and the copolymerization take place competitively, a 1,4-diradical is very likely to be the common intermediate. The stereoselectivity of the cycloaddition does not completely rule out the stepwise mechanism. The singlet 1,4-diradical has a very short lifetime and may be constrained from rotation by the presence of bulky aromatic moieties. So if it cyclizes faster than it can rotate about the single bond, the predominant products will retain the initial structure. Furthermore, various degrees of stereoselectivity have been observed in some photocycloaddition systems and were attributed to the existence of several exciplex minima. The stepwise reaction may afford an alternative explanation.

A detailed description of the singlet exciplex decay channels remains an intriguing problem for both the organic chemists and polymer chemists. Polymerization as a powerful way of trapping and characterizing the intermediates will surely play a very important role in clarifying this subject, just as in the spontaneous thermal reactions.

4.3 Tetramethylene Diradicals as the Intermediates in Initiating Copolymerization of Donor/Acceptor Olefins

Raetzsch and Shirota, two notable photochemists in this area, studied several copolymerization systems and both proposed radical-ions as the initiating intermediates [14, 24, 26]. However, insufficiently detailed schemes and evidence were given to support this interpretation.

We propose an alternative mechanism: bond formation to tetramethylene intermediate. The results which support this proposal come from not only our own experiments, but also from Raetzsch's and Shirota's data.

Raetzsch observed the increase of molecular weight of the copolymer with the monomer conversion in St/MAn [15]. This is excellent evidence for the diradical initiation and propagation; termination in a diradical system consists of coupling of chains with a radical on each end, resulting in molecular weight increase. We also obtained similar results in 2-VN/FN and 2-VN/MAn system [37]. It is noteworthy that the "diradical polymerization" cannot be "pure" and is always accompanied by monoradical side-reactions. Chain transfer, disportionation of radicals, and deactivation of one end of the diradical may produce monoradicals, which play the same role as monofunctional compounds in polycondensation. The increase of polymer molecular weight caused by diradical coupling will be hindered or suppressed. So it is not surprising that many systems show only a slight increase of molecular weight with time while others do not show it at all. Dimonie et al. [120] derived a functional relationship between "instantaneous" polymerization degree and conversion in normal radical polymerization, in which molecular weight (MW) decreases as the monomer is consumed with time. So even if the MW keeps constant during a polymerization, a diradical propagation may not be ruled out.

Another important variable is the dependence of the yields of copolymer and oligomers on the olefin concentration. In many systems [38, 121], it was found

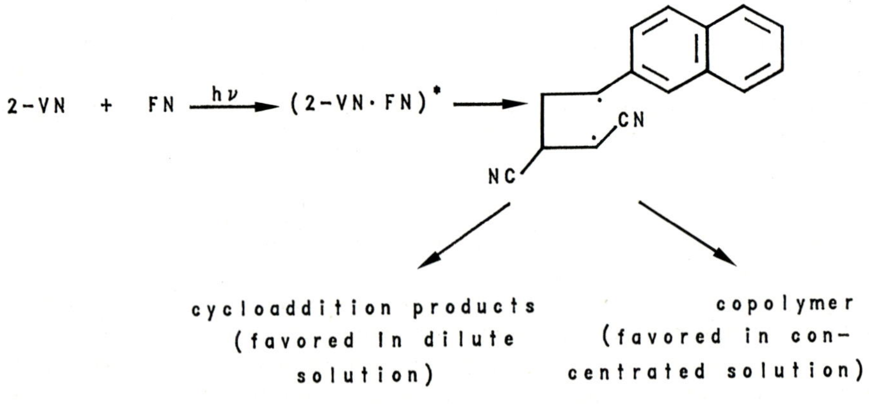

$$2\text{-VN} + \text{FN} \xrightarrow{h\nu} (2\text{-VN}\cdot\text{FN})^* \longrightarrow \text{[cyclobutane intermediate]}$$

cycloaddition products (favored in dilute solution)

copolymer (favored in concentrated solution)

(41)

that the oligomer yield increases and the copolymer yield decreases as functions of decreasing monomer concentrations. This suggests a common intermediate for both the two products. For example, in the photoreaction of 2VN-FN system, the following scheme was proposed (see (41)).

We also studied the photoreaction of St/AN system which has no ground state complex. St is excited first, and then forms an excimer with another St molecule or forms an exciplex with AN. The following reaction scheme was proposed:

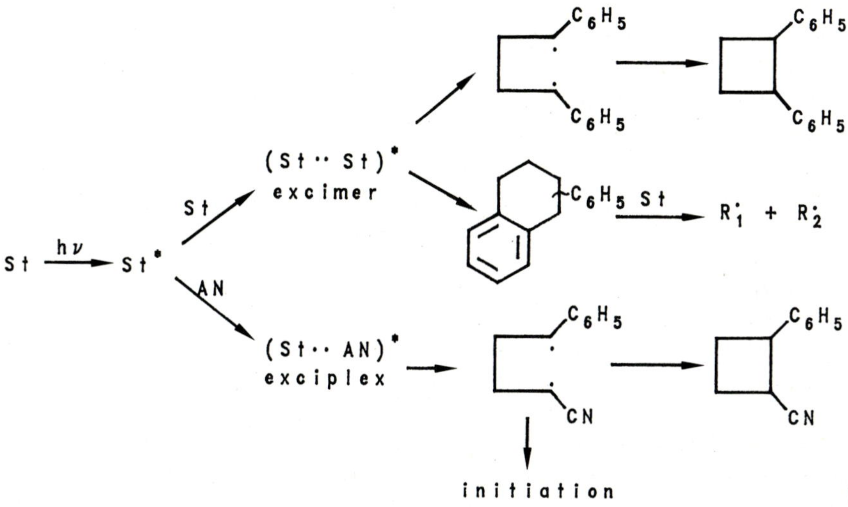

$$
\text{St} \xrightarrow{h\nu} \text{St}^*
\begin{cases}
\text{St} \to (\text{St} \cdots \text{St})^* \text{ excimer} \\
\text{AN} \to (\text{St} \cdots \text{AN})^* \text{ exciplex}
\end{cases}
\quad \xrightarrow{\text{C}_6\text{H}_5\text{St}} R_1^{\cdot} + R_2^{\cdot}
$$

initiation (42)

The St/St diradical is mainly responsible for the cyclization, while the St/AN diradical, which was trapped by TEMPO, is effective for the polymerization initiation. When the photocopolymerization was carried out in the presence of benzophenone, higher MW and remarkable MW increase with time was observed, since in this case triplet benzophenone transferred its energy to St, so the triplet exciplex and then the triplet diradical were produced. A triplet 1,4-diradical has a longer lifetime than a singlet: this favors the transformation from gauche to trans conformation, and thus favors further monomer addition. Turro has shown that a singlet diradical has a "fight" geometry for which the free valences are close together in space, whereas a triplet diradical will favor a loose geometry for which the free valence are as far apart as possible [108].

5 Conclusion

The initiation mechanism of photo-induced charge-transfer copolymerization of donor-acceptor monomer pairs is clarified by integrating the results from organic chemistry and polymer chemistry. A reasonable suggestion for the initiating species in certain cases may be a tetramethylene 1,4-diradical. The excited complex of the donor/acceptor monomers undergoes multiple follow-up reactions to produce the tetramethylene intermediates. The singlet state exciplex might collapse to a singlet 1,4-diradical in addition to the concerted cycloadition. Triplet 1,4-diradicals are also known. Bond formation can take place in radical-ion reactions, too, producing a 1,4-cationic radical tetramethylene. The 1,4-diradical tetramethylene may be the key intermediate in photocycloaddition as well as photocopolymerization. Although some aspects remain unclear or even suspicious and the initiation mechanism might be further complicated by possible non-bond-formation reactions, such as proton transfer and reaction between monomer and solvent molecules [13], they will be clarified as long as we follow the available research clues. "Bond-Formation Initiation" could serve as a unifying link between the spontaneous thermal reactions and photoreactions of donor and acceptors, as well as the small molecule photochemistry (cycloadditions) and photopolymerizations of donors and acceptors.

Acknowledgement: We are deeply indebted to the National Nature Science Foundation of China and to the National Science Foundation, Division of Materials Research for support of this research. We also wish to thank Professor R. A. Caldwell for a critical reading of the manuscript and Dr. Samir Farid and Professor J. Kochi for very helpful discussions.

6 References

1. Li T, Cao W, Feng X (1987) Scientia Sinica (B) 7:685
2. Bouas-Laurent H, Castellan A, Desvergne J-P (1980) Pure Appl. Chem. 52:2633
3. Cowan DO, Drisko RL (1976) Elements of organic photochemistry. Plenum, New York, Chap. 2
4. McCullough JJ, Wu WS, Huang CW (1972) J. Chem. Soc. Perkin Trans. 2:371
5. Shirota Y, Mikawa H (1977–1978) J. Macromol. Sci.-Rev. Macromol. Chem. C16:129
6. Shirota Y, Mikawa H (1985) Mol. Cryst. Liq. Cryst. 126:43
7. Shirota Y (1985) in: Mark HF, Kroschwitz, JI (eds) Encyclopedia of polymer science and engineering. 2nd ed, Wiley, New York, 3:327
8. Irie M, Hayashi K (1975) Progress in Polymer Science in Japan 8:105
9. Li T, Cao W, Feng X (1989) J. Macromol. Sci.-Rev. Macromol. C29:153
10. Bowman RM, Calvo C, McCullough JJ, Miller RC, Singh I (1973) Can. J. Chem. 51:1060
11. Akhtar IA, McCullough JJ (1981) J. Org. Chem. 46:1447
12. Barton J, Capek I, Hrdlovic P (1975) J. Polym. Sci., Polym. Chem. Ed. 13:2671
13. Capek I, Barton J (1975) J. Polym. Sci., Polym. Chem. Ed. 13:2691
14. Raetzsch M (1975) Progr. Polym. Sci. 13:277
15. Barton J, Capek I, Arnold M, Raetzsch M (1980) Makromol. Chem. 181:241
16. Raetzsch M, Chomiakov K (1980) Acta Polym. 30:577
17. Ottenbrite RM, Jones J (1985) ACS Polymer Prepr. 26:128
18. Gaylord NG, Maiti S, Dixit SS (1972) J. Macromol. Sci.-Chem. 6:1521
19. Hallensleben ML (1973) Eur. Polym. J. 9:228
20. Li T, Pan J, Zhang Z (1989) Makromol. Chem. 190:1391
21. Hallensleben ML (1970) Makromol. Chem. 144:267
22. Li X, Chen G, Li S, Qin A, Yu T (1988) Makromol. Chem. Rapid. Commun. 9:195
23. Lee C, Hall HK Jr. (1989) Macromolecules 22:21
24. Oh S-C, Yamaguchi K, Shirota Y (1987) Polym. Bull. 18:99
25. Barb WG (1953) Trans. Faraday Soc. 49:143

26. Yamaguchi K, On S-C, Shirota Y (1986) Chem. Lett. 1445
27. Oshiro Y, Shirota Y, Mikawa H (1973) Chem. Lett. 1299
28. Gaylord NG, Dixit SS (1971) J. Polym. Sci. (B) 8:823
29. Tazuke S, Okamura S (1969) J. Polym. Sci. A-I 7:715
30. Furukawa J, Kobayashi E, Iseda Y (1970) J. Polym. Sci. (B) 8:47
31. Ellinger LP (1964) Polymer 5:559
32. Itoh M (1974) J. Am. Chem. Soc. 96:7390
33. Shirota Y, Tsushi I, Mikawa H (1974) Bull. Chem. Soc. Jpn. 47:991
34. Tada K, Shirota Y, Mikawa H (1972) J. Polym. Sci., Polym. Lett. Ed 10:69
35. Lewis FD (1979) Acc. Chem. Res. 12:153
36. Green BS, Rejto M, Johnson DE, Hoyle CE, Ho TI, McCoy F, Simpson JT, Lewis FD (1979) J. Am. Chem. Soc. 101:3325
37. Li T, Luo B, Chu G, Hall HK Jr. J. Pol. Sci. Pol. Chem. Ed. (in press)
38. Li T, Padias, AB, Hall HK Jr. Macromolecules (in press).
39. Lewis FD, Devoe RJ (1982) Tetrahedron 38:1069
40. DeMayo P (1971) Acc. Chem. Res. 4:41
41. Wamser CC, Hammond GS, Chang CT, Baylor C (1970) J. Am. Chem. Soc. 92:6362
42. Lopp IG, Hendren RW, Wildes PD, Whitten DG (1970) J. Am. Chem. Soc. 92:6440
43. Roy JK, Whitten DG (1971) J. Am. Chem. Soc. 93:7094
44. Forster Th (1969) Angew. Chem. Int. Ed. Engl. 8:333
45. Corey GJ, Bass JD, Le Mahieu R, Mitra RB (1964) J. Am. Chem. Soc. 86:5870
46. Coxon JM, Halton B (1987) Organic photochemistry. 2nd ed. Cambridge University Press, London, p 142
47. Beens H, Weller A (1968) Chem. Phys. Lett. 2:140
48. Roy JK, Whitten DG (1971) J. Am. Chem. Soc. 93:7093
49. Caldwell RA, Creed D, Ohta H (1975) J. Am. Chem. Soc. 97:3246
50. Saltiel J, Townsend DG, Watson BD, Shannon P (1975) J. Am. Chem. Soc. 97:5688
51. Creed D, Caldwell RA, Ohta H, DeMarco DC (1977) J. Am. Chem. Soc. 99:277
52. Masahara H, Mataga N, Yoshida M, Tatemitsu H, Sakata Y, Misumi S (1977) J. Phys. Chem. 81:879
53. Lim BT, Okajima S, Chandra AK, Lim ZC (1987) J. Chem. Phys. 77:3902
54. Yang N-C, Shou H, Wang T, Masnovi J (1980) J. Am. Chem. Soc. 102:6652
55. Lewis FD, Johnson DE (1978) J. Am. Chem. Soc. 100:983
56. Green BS, Rejto M, Johnson DE, Hoyle CE, Simpson JT, Correa PE, Ho TI, McCoy F, Lewis FD (1979) J. Am. Chem. Soc. 101:3325
57. Calhoun GC, Schuster GB (1986) J. Am. Chem. Soc. 108:8021
58. Akbulut N, Hartsough D, Kim J-I, Schuster GB (1989) J. Org. Chem. 54:2549
59. Greed D, Caldwell RA (1974) J. Am. Chem. Soc. 96:7396
60. Caldwell RA, Smith L (1974) J. Am. Chem. Soc. 96:2994
61. Caldwell RA, Greed D (1980) Acc. Chem. Res. 13:45
62. Caldwell RA, Chali NI, Chien C-K, DeMarco D, Smith L (1978) J. Am. Chem. Soc. 100:2857
63. Michl J (1977) Photochem. Photobiol. 25:141
64. Gerhartz W, Poshusta RD, Michl J (1976) J. Am. Chem. Soc. 98:6427
65. Mitters SL, Farad S (1982) Acc. Chem. Res. 15:80
66. Ellinger LP (1964) Polymer 5:599
67. Ledwith A (1972) Acc. Chem. Res. 5:133
68. Ledwith A (1985) in: Gordon M, Ware WR (eds) The Exciplex. Academic Press, New York, p 209
69. Kuwata S, Shingemitsu Y, Odaira Y (1973) J. Org. Chem. 38:3803
70. Evans TR, Wake RW, Jeanicke O (1975) in: Gordon M, Ware WR (eds) The Exciplex. Academic Press, New York, p 345
71. Asanuma T, Yamamoto M, Nishijima Y (1975) J. Chem. Soc. Chem. Commun. 608
72. Kojima M, Sakuragi H, Tolumaru K (1981) Tetrahedron Lett. 22:2889

73. Farid S, Hartmann SE, Evans TR (1975) in: Gordon M, Ware WR (eds) The Exciplex. Academic, New York, p 327
74. Mattes S, Farid S (1983) in: Padwa A (ed) Organic photochemistry Marcel Dekker, New York, Vol 6, p 233
75. Mattes SL, Luss HR, Farid S (1983) J. Phys. Chem. 87:4779
76. Neurteufel RA, Arnold DR (1973) J. Am. Chem. Soc. 95:4080
77. Majima T, Pac C, Nakasone A, Sakurai H (1981) J. Am. Chem. Soc. 103:4499
78. Pac C (1986) Pure Appl. Chem. 58:1249
79. Mizuno K, Kangano H, Kasuga T, Otsuji Y (1983) Chem. Lett. 133
80. Yamamoto M, Asanuma T, Nishijima Y (1975) J. Chem. Soc. Chem. Commun. 53
81. Mattary J (1987) Angew. Chem. Int. Ed. Engl. 26:825
82. McCullough JJ, Miller RC, Fung D, Wu W-S (1975) J. Am. Chem. Soc. 97:5492
83. McCullough JJ, Miller RG, Wu W-S (1977) Can. J. Chem. 55:2909
84. Lewis FD (1986) Acc. Chem. Res. 19:401
85. Bowman RM, Chamberlain TR, Huang CW, McCullough JJ (1974) J. Am. Chem. Soc. 96:692
86. Hall HK Jr. (1983) Angew. Chem. Int. Ed. Engl. 22:440
87. Gotoh T, Padias AB, Hall HK Jr. (1986) J. Am. Chem. Soc. 108:4902
88. Hall HK Jr., Padias AB, Pandya A, Tanaka H (1987) Macromolecules 20:247
89. Huisgen R (1977) Accts. Chem. Rev. 10:149
90. Rasoul HAA, Hall HK Jr. (1982) J. Org. Chem. 47:2080
91. Kushibiki N, Irie M, Hayashi K (1975) J. Polym. Sci., Polym. Chem. Ed. 13:77
92. Platz MS (1982) in: Borden WT (ed) Diradicals. Wiley, New York, p 195
93. Hamify H, Scaiano JC (1975) J. Photochem. 4:229
94. Arnold DR (1968) Adv. Photochem. 6:301
95. Jones G II (1981) in: Padwa A (ed) Organic photochemistry. Vol 5, Marcel Dekker, New York, p 1
96. Wilson RM (1985) in: Padwa A (ed) Organic photochemistry. Vol 9, Marcel Dekker, New York, p 340
97. Yang NC, Eisenhardt W (1971) J. Am. Chem. Soc. 93:1277
98. Pagni RM, Burnett MN, Dodd JR (1977) J. Am. Chem. Soc. 99:1972
99. Engel PS (1980) Chem. Rev. 80:99
100. Wilson RM, Geiser F (1978) J. Am. Chem. Soc. 100:2225
101. Adam W, Hannemann K, Wilson RM (1984) J. Am. Chem. Soc. 106:7646
102. Loutty RO, DeMayo P (1977) J. Am. Chem. Soc. 99:3559
103. Becker D, Nagler M, Hirsch S, Ramun J (1983) J. Chem. Soc. Chem. Commun. 371
104. Maruyama K, Oksuki T, Tai S (1985) J. Org. Chem. 50:52
105. Horspool WM (1976) Aspects of organic photochemistry. Academic Press, London, Chap 4
106. Lewis FD, Kojima M (1988) J. Am. Chem. Soc. 110:8660
107. Caldwell RA (1978) J. Am. Chem. Soc. 100:2857
108. Turro NJ (1978) Modern molecular photochemistry. Benjamin/Cummings Publ Co, Chap 11
109. Lewis FD, DeVoe RJ, MacBlane DB (1982) J. Org. Chem. 47:1392
110. Lewis FD, Hirsch RH (1976) J. Am. Chem. Soc. 98:5914
111. Hilinski EF, Masnovi JM, Kochi JK, Rentzepis PM (1983) J. Am. Chem. Soc. 105:6167
112. Hilinski EF, Masnovi JM, Kochi JK, Rentzepis PM (1984) J. Am. Chem. Soc. 106:8071
113. Masnovi JM, Kochi JK, (1985) J. Org. Chem. 50:5245
114. Okada T, Kida K, Mataga N (1982) Chem. Phys. Lett. 88:157
115. Wender PA (1986) in: Coyle JD (ed) Photochemistry in organic synthesis. The Royal Society of Chemistry Burlington House, Chap 9
116. Woodward RB, Hoffmann R (1970) The conservation of orbital symmetry. Academic Press, New York
117. Mitters SL, Farid S (1982) Acc. Chem. Res. 15:80

118. Chapman DL, Lura RD, Owens RM, Plank ED, Shin SC, Arnold DR, Gillis LB (1972) Can. J. Chem. 50:1984
119. Green BS, Rejto M, Johnson DE, Hoyle CE, Ho TI, McCoy F, Simpson JT, Lewis FD (1979) J. Am. Chem. Soc. 101:3325
120. Dimonie M, Oplescu CR, Hubca Gh (1976) Rev. Roum. Chim. 21:763
121. Li T, Padias AB, Hall HK Jr Macromolecules (in press)

Editor: J. P. Kennedy
Received October 25, 1989

Polyfunctional Cyanate Monomers as Components of Polymer Systems

Piotr Penczek and Wiesława Kamińska
Industrial Chemistry Research Institute, ul. Rydygiera 8, 01-793 Warsaw, Poland

Cyclotrimerization of cyanate groups in 2,2-bis(4-cyanatophenyl)propane and other polyfunctional cyanate monomers results in formation of polymer networks with triazine rings. Cyclotrimerization of cyanates in systems containing thermoplastic polymers yields semi-IPNs (Interpenetrating Polymer Networks) with elevated heat deflection temperatures. Numerous systems consisting of a polyfunctional cyanate monomer and a bismaleimide are described. Such systems can contain, moreover, a third crosslinkable monomer, e.g. an allyl ester or an epoxy resin. Two-component systems consisting of a cyanate monomer and an epoxy resin with corresponding curing agent have been developed. Some other resins and monomers, e.g. phenolics, acrylics and ester imides, were also used in compositions with cyanate monomers. Rubber-flexibilized 2,2-bis(cyanatophenyl)propane polymers have been patented. The authors' investigations on IPNS, which were obtained by simultaneous crosslinking of 2,2-bis(4-cyanatophenyl)propane and unsaturated polyester resin, are described.

List of Abbreviations 42

1 Introduction 42

2 Polycyclotrimerization of Dicyanates 43

3 Copolymers of Dicyanates with Bifunctional Monomers 45

4 Thermoplastic Polymers and Cyanate Monomers 47

5 Heat Resistant Polymers from Polyfunctional Cyanates and Maleimides . . . 48

6 Epoxide Resin Compositions with Polyfunctional Cyanates 49

7 Three-Component Cyanate/Epoxide/Maleimide Systems 53

8 Polyfunctional Unsaturated Monomers and Cyanates 55

9 Compositions of Polyesters, Polyurethanes, Modified Phenolic Resins and Related Materials with Dicyanates 57

10 Rubbers Modified with Dicyanates 57

11 References 58

Advances in Polymer Sciences 97
© Springer-Verlag Berlin Heidelberg 1990

List of Abbreviations

BPA	—	Bisphenol A; 2,2-bis(4-hydroxyphenyl)propane
BPA/DC	—	Bisphenol A dicyanate; 2,2-bis(4-cyanatophenyl)propane
BMI	—	bismaleimide; bis(4-maleimidophenyl)methane
ECH	—	epichlorohydrin
DABCO	—	triethylenediamine; 1,4-diazabicyclo[2.2.2]octane
AIBN	—	azobisisobutyronitrile
PPO	—	polyphenylene oxide; poly(oxy-2,6-dimethylphenylene)
UPR	—	unsaturated polyester resin
IPN	—	Interpenetrating Polymer Network
T_g	—	Glass Transition Temperature

1 Introduction

Aromatic cyanates are synthesized by the condensation of phenols with ClCN and a tertiary amine [1]. The most important dicyanate is 2,2-bis(4-cyanatophenyl) propane or Bisphenol A dicyanate [2] (Scheme 1).

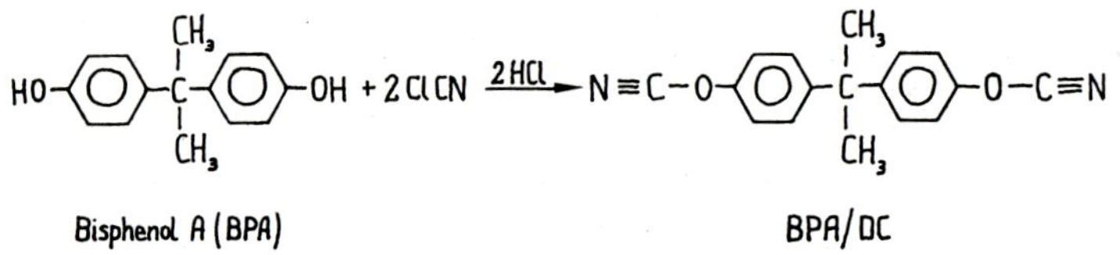

$$HO-\underset{Bisphenol\ A\ (BPA)}{\bigcirc-\underset{\underset{CH_3}{|}}{\overset{\overset{CH_3}{|}}{C}}-\bigcirc-OH} + 2\,ClCN \xrightarrow{2HCl} \underset{BPA/DC}{N\equiv C-O-\bigcirc-\underset{\underset{CH_3}{|}}{\overset{\overset{CH_3}{|}}{C}}-\bigcirc-O-C\equiv N}$$

Scheme 1

Aromatic cyanates trimerize on heating in the presence of catalysts, thus forming triaryl cyanurates (Scheme 2)

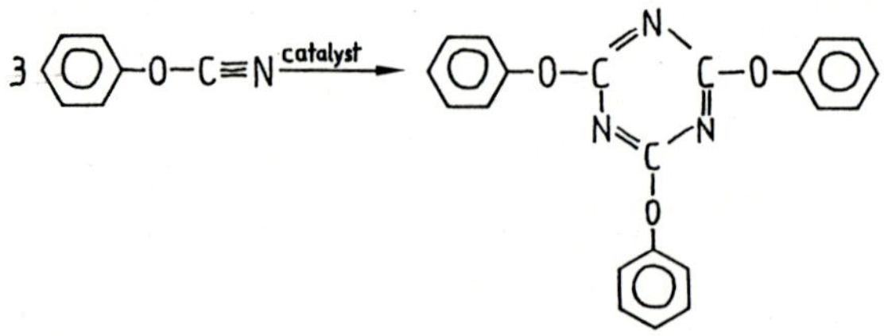

$$3\,\bigcirc-O-C\equiv N \xrightarrow{catalyst} \text{triaryl cyanurate}$$

Scheme 2

As a result of the trimerization of dicyanates, crosslinked polymers consisting of benzene and triazine rings are formed (Scheme 3).

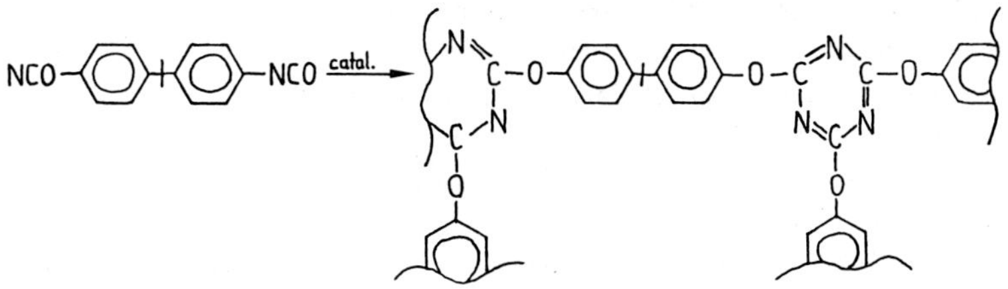

Scheme 3

The crosslinked triazine (or cyanurate) polymers distinguish themselves by high glass temperatures and improved thermal stability.

Crosslinked polytriazines (polycyanurates) with still higher crosslinking density are obtained from polycyanates based on novolak resins (Scheme 4)

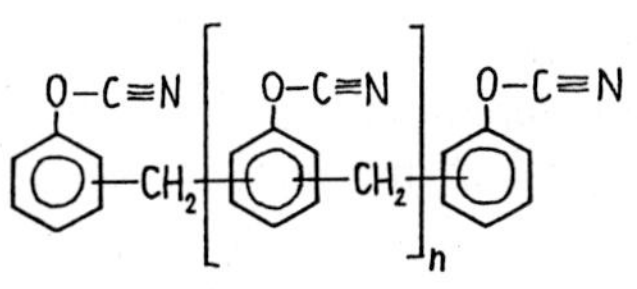

Scheme 4

If flexibilization of the very rigid polytriazine network is desired, monoaryl cyanates are added to the dicyanate monomers. In some patents the use of *p*-isopropylphenyl (cumyl) cyanate (Scheme 5) is mentioned.

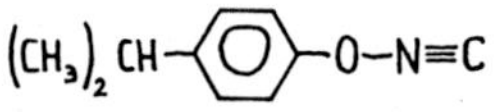

Scheme 5

2 Polycyclotrimerization of Dicyanates

Among the patents and patent applications concerning the uses of dicyanates in polymer technology, the information prevails on systems which consist of dicyanates and other monomers and polymers. Although such systems are the object of this article, the polymer materials obtained solely from aromatic dicyanates are reviewed shortly, too, in order to present a more complete set of information.

A comprehensive review on the synthesis and the polycyclotrimerization of cyanates was published in the USSR in 1977 (with 160 references) [3]. More than 50 dicyanates were described, including compounds with cyanate groups attached to single, bridged or condensed aromatic rings. Very high thermal stability is shown by cyanate polymers with carborane bridges between benzene rings. Dicyanates with aromatic oligoether, ketone-ether and sulfone-ether as well as oligodimethylsiloxane or oligomethylphenylsiloxane backbones were also mentioned. Aliphatic dicyanates do not tend to premature cyclotrimerization if they contain perfluoromethylene chains. Polymers with pendant cyanatophenyl substituents were reviewed.

As the cyclotrimerization catalysts, cationic ($SnCl_4$, $ZnCl_2$), anionic (triethylamine, pyridine, triphenylphosphine) and coordination-ionic (Cr^{+3} acetylacetonate, with water as cocatalyst) ones can be used [4]; $AlCl_3$ was used as well [5].

Kinetics of the cyclotrimerization was investigated using IR spectra, the disappearance of the 2240–2285 cm^{-1} stretching band (C≡N) in particular [6, 7]. A similar method (2230 and 2270 cm^{-1}, with the CH stretching band in the region 3000–3100 cm^{-1} as internal standard) was used for the determination of OCN conversion in both soluble and gelled BPA/DC cyclotrimerizates during the synthesis of BPA/DC prepolymers [8]. Thermal effects of the cyclopolymerization were also taken into account [7, 9, 10]. For the cyclotrimerization without catalysts under isothermal conditions and with a constant heating rate, activation energy of 80 kJ/mole for the first order brutto kinetics was found. Enthalpy of the polycyclotrimerization is 110 kJ/mole OCN [11]. The OCN conversion can be thus calculated from the DSC measurements; the results agree with those from IR spectroscopy [11, 12].

Individual oligomeric BPA/DC cyclotrimerizates were separated by the HPLC gradient-elution technique in THF/H_2O [13]. The step-growth mechanism of the polycyclotrimerization was supported by quantitative analysis of the results.

Catalyst-free aromatic cyanates undergo cyclotrimerization only if they contain phenolic substances and water. Phenolic substances (non-reacted BPA and BPA monocyanate) are usually present in BPA/DC. Water evaporates gradually on

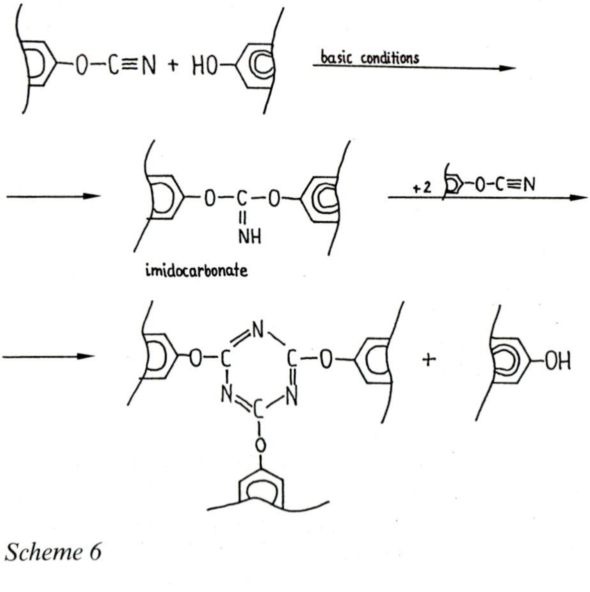

Scheme 6

heating and the reaction continues according to Scheme 6, with consecutive phenol addition and abstraction [12]. This mechanism was supported by the experiments on BPA/DC mixtures with BPA monocyanate or imidocarbonate added. Cyclotrimerization kinetics was investigated and mathematical modelling was carried out for the BPA/DC-phenolic OH system [12]. As it is usual, the reaction is slower after T_g reaches the reaction temperature.

High mechanical strength, very good adhesion and outstanding thermal stability of the polycyanurates obtained from various dicyanates (those with voluminous bridges between the aromatic rings in particular) were pointed out [6]. The chemical structure of dicyanate monomers (except those with carborane bridges) has little effect on the thermal decomposition of the polymers (beginning of decomposition in air and in an inert atmosphere at 385–405 °C and 390–450 °C as well as second exothermic effect at 490–580 °C and 500–600 °C, respectively) [14]. Investigation of the thermal decomposition products has shown that the weakest part of the polymer is the C—O—C bond [15].

Two papers published on BPA/DC appeared in the mid 1960s [16, 17]. Comprehensive information on the BPA/DC based thermostable polymer materials (Triazine A Resins, TA-Harze) was then published [18].

To obtain a heat and water resistant crosslinked polymer, Cu naphthenate and an alcohol (e.g. benzyl alcohol) were added to BPA/DC as a catalyst; epoxide resins (cf. Sect. 6) can be used as well [19]. The use of Zn acetate together with dicumyl peroxide was also mentioned [20].

Transparent heat resistant coatings were formed by heating of a mixture of BPA/DC, 1,1,1-tris(4-cyanatophenyl)ethane, nonylphenol and Zn naphthenate [21]. The crosslinked polymer from BPA/DC, maleic anhydride, p-toluenesulfonic acid, Zn acetate and DABCO has high T_g [22]. BPA/DC monomer can be replaced by the corresponding prepolymer with p-toluenesulfonic acid monohydrate and Zn acetate. The composition obtained is processed as a molding compound [23].

BPA/DC-based composition served as a polymer matrix in carbon fiber composites [24]. Metal powder or fiber filled molding compounds were also described. The molding compounds, which can contain, moreover, numerous other components, bismaleimides (cf. Sect. 5) in particular, are used for the manufacturing of heat conductive molds for injection molding [25].

3 Copolymers of Dicyanates with Bifunctional Monomers

Polyaddition of dicyanates with various bifunctional monomers leads to linear, branched or crosslinked polymer [3].

The reaction with cyclic anhydrides of dicarboxylic acids results in the formation of imidocarbamates [26–28] (Scheme 7).

From BPA/DC and pyromellitic dianhydride, linear polyimidocarbamates were synthesized [27–29]. For the same purpose, various dicyanates and tetracarboxylic acid dianhydrides were used [29–31]. If one of the starting monomers has a functionality higher than 2, crosslinked polymers are obtained,

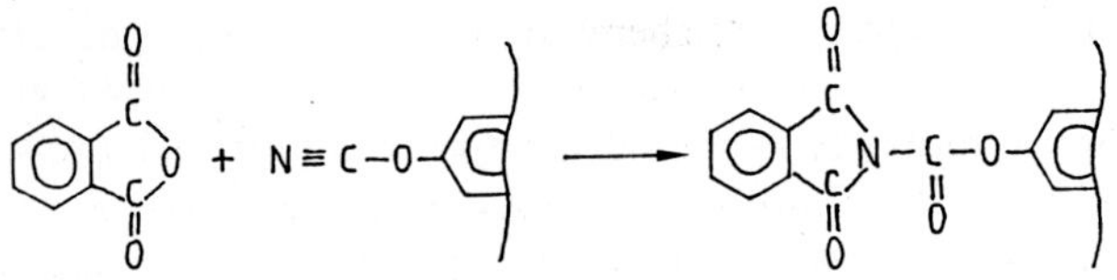

Scheme 7

Dicyanates react with diisocyanates. The polyaddition results in the formation of crosslinked polymers with cyanurate and isocyanurate rings [32]. The polymers have high deformation and decomposition temperature: in the case of a crosslinked copolymer from 9,9-bis(4-cyanatophenyl)fluorene the corresponding values are 480 °C and 440 °C (in air), respectively.

Linear polymers were obtained by the polyaddition of dicyanates with the compounds, which contain active protons: primary and secondary amines, glycols and bisphenols. The reaction with diamines given polyisoureas [30, 31] (Scheme 8).

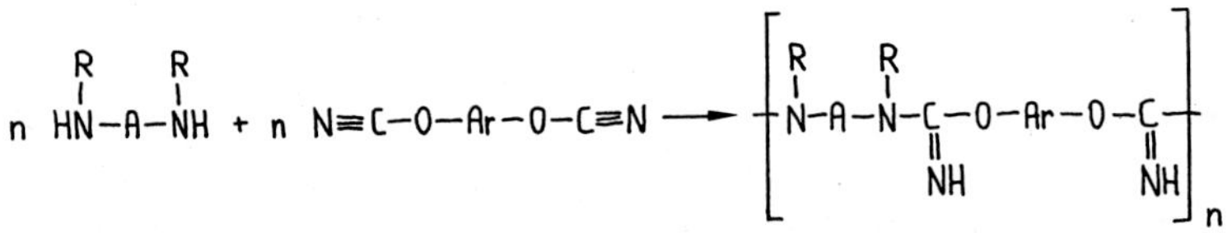

Scheme 8

The polyaddition with glycols and bisphenols results in the formation of polyiminocarbonates (Scheme 9). The reaction proceeds in the presence of strong alkaline catalysts only [29, 33].

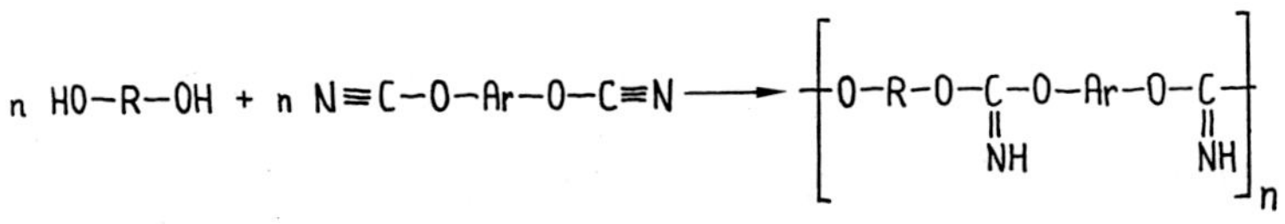

Scheme 9

The linear addition copolymers of dicyanates are synthesized at equimolecular ratio of the comonomers. If one of the comonomers is used in excess, lower-molecular-weight polymers with corresponding end groups are obtained, which can further react with chain elongation agents. Terminal cyanate groups can be cyclotrimerized to obtain crosslinked polymers.

4 Thermoplastic Polymers and Cyanate Monomers

An increase in heat deflection temperature of some thermoplastic polymers can be achieved by the addition of polyfunctional aromatic cyanates (BPA/DC in particular) and trimerization catalysts. A rigid network is formed as a result of the cyanate trimerization. The polymer material consists of a linear polymer and a crosslinked network and belongs to the class of semi-IPNs (semi-Interpenetrating Polymer Networks); the corresponding classification is given in [34–37].

The cyanates are mixed with thermoplastic polymers. They can be also dissolved in a monomer and then a simultaneous polymerization is carried out. So-called SINs (Simultaneous Interpenetrating Networks) are obtained.

Plastic material for printed circuits was obtained from a mixture of polycarbonate or polyethersulfone with BPA/DC [38].

For the same purpose, a composition consisting of Bisphenol A copoly (carbonate-terephthalate) and BPA/DC was reinforced with polyamide fibers. After crosslinking at 270 °C, a glass transition temperature of 212 °C was obtained [39]. The same dicyanate monomer was added to polycarbonate in order to decrease the brittleness [40].

BPA/DC was added to polyoxymethylene in order to increase the adhesion to glass fibers [41].

Laminates for printed circuits were obtained from filled polypropylene with <5% BPA/DC added [42].

A series of Japanese patent applications concerns addition of BPA/DC to poly(2,6-dimethylphenylene oxide) (PPO). Electroinsulating coatings with elevated heat resistance were obtained from PPO and BPA/DC [43]. As a rule, maleimide monomers are used together with BPA/DC. According to [44], the following reaction (Scheme 10) between BPA/DC and maleimide takes place:

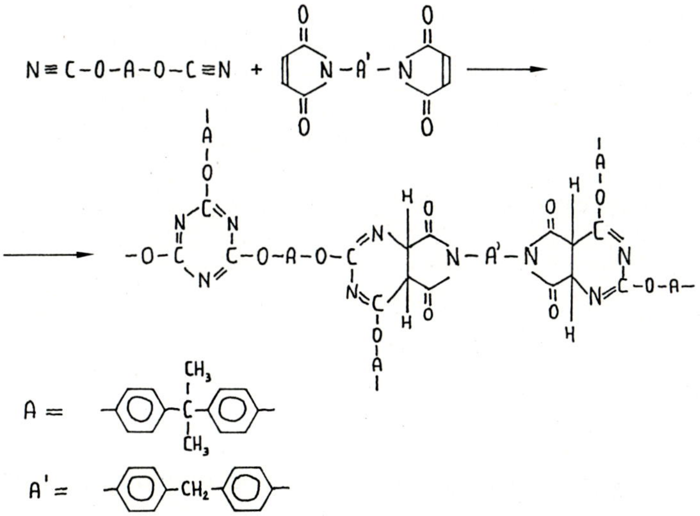

Scheme 10

A thermohardening composition for glass fiber reinforced plastics consists of PPO, BPA/DC and a bismaleimide [45]. Still higher heat resistance is achieved, if the polyfunctional maleimide, based on aniline-formaldehyde condensation products (Scheme 11), is used [46].

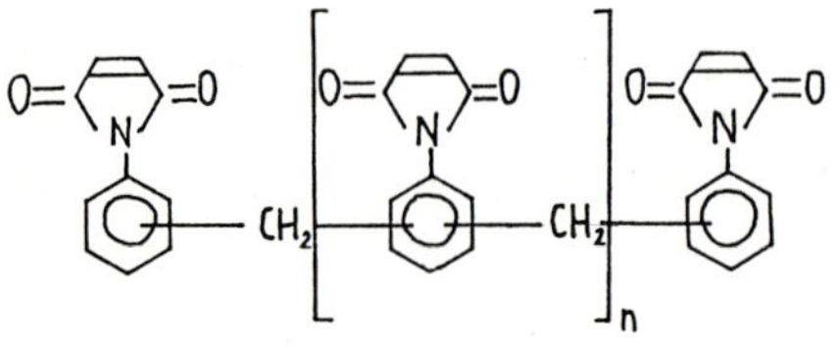

Scheme 11

Solvent resistant laminates for printed circuits were manufactured by coating of copper foil with a solution of PPO, BPA/DC, bis(4-maleimidophenyl) ether and Zn octoate in toluene; the coated foil was laminated with PPO-impregnated glass fabric [47]. Similar result was achieved by the modification of PPO with polyfunctional cyanates or maleimides, liquid polybutadiene and a polymerization catalyst [48]. A solvent and heat resistant composition for printed circuits consists of copoly [(2,6-dimethylphenylene)-(2,3,6-trimethylphenylene)]oxide, maleic anhydride grafted poly-1,2-butadiene, bis(4-maleimidophenyl)methane, BPA/DC and toluene. BPA/DC prepolymer may be used instead of the monomer [49].

BPA/DC was compounded with high-temperature resistant ladder and semi-ladder polymers. A composition, which contained BPA/DC, a polyimide resin, organic Zn salt and benzyldimethylamine in N-methylpyrrolidone solution was described. The binder was destined for the manufacture of copper clad laminates. T_g of the crosslinked material was 265 °C [50]. Another example is a binder obtained from BPA/DC and polyphenylquinoxaline in chloroform [51].

Polystyrene modified dicyanate polymer was obtained by simultaneous polymerization of styrene monomer and BPA/DC. The handling of low viscous dicyanate solution in styrene is more convenient that the processing of crystalline BPA/DC or high-viscous dicyanate prepolymers [52]. A similar patent specification describes the polymerization of a mixture of styrene, BPA/DC and p-isopropenyl cyanate; BMI (cf. Sect. 5) was also used [53].

Acrylic monomers: glycidyl methacrylate together with an epoxide resin [54], 2-hydroxyethyl methacrylate [55] and 2-ethylhexyl acrylate [56] containing dissolved BPA/DC were heated to obtained prepolymers, which were further polymerized; a peroxide initiator or UV irradiation was applied.

5 Heat Resistant Polymers from Polyfunctional Cyanates and Maleimides

The reaction of dicyanates and bismaleimides leads to crosslinked polymers (Scheme 10). Separate trimerization of the dicyanate and radical polymerization of bismaleimide can also occur, thus forming an IPN.

Heating of a mixture of BPA/DC and BMI at 100–160 °C results in a formation of prepolymers, known as BT Resins of Mitsubishi Gas Chemical Co. [57]. A broad assortment of BT Resins has been developed. Some of the BT Resins are modified with epoxide resins, unsaturated monomers etc.

The resins contain 10–60% BMI; the molecular weight amounts to 290–2300 and the melting point is in the range of 30–130 °C. Glass transition temperatures of the crosslinked products are 230–290 °C. Maximum working temperature is 180–210 °C. The BT Resins are used for the impregnation of electric motor coil windings, as engineering materials in aircraft, in reinforced plastics (in copper clad laminates in particular), injection molding powders and powder coatings.

Numerous cyanate/maleimide compositions and the applications thereof have been described in Japanese patent applications.

A binder for copper clad laminates contains BPA/DC, bis(4-maleimidophenyl)-methane (BMI) and Zn acetate as trimerization catalyst [58]. Heat resistant coating for steel sheets is based on BPA/DC and bis(4-maleimidophenyl) ether [59]. Good storage stability, short curing time and high thermal resistance has a binder obtained from BPA/DC, BMI as well as metal chelates and H_2O_2 [60]. A composition for copper clad laminates was obtained by heating of BPA/DC with a bismaleimide at 120 °C and then adding Zn acetate and triethylenediamine [61].

Instead of BPA/DC monomer, the corresponding prepolymer was heated with BMI to obtain a binder [62]. In a similar composition, p-toluenesulfonic acid was used as the cyanate/maleimide prereaction catalyst. For the crosslinking stage Zn octoate was added [63]. Both BPA/DC and BMI can be used in the form of prepolymers. The binder for copper clad laminates contains, moreover, a saturated polyester, Zn octoate and powdered fused quartz [64].

Thermoplastic preimpregnates were obtained by reacting of BPA/DC with BMI and an aromatic diamine [65]. The well-known addition of amino groups to maleimide double bonds results in the formation of oligoaspartimides (Scheme 12).

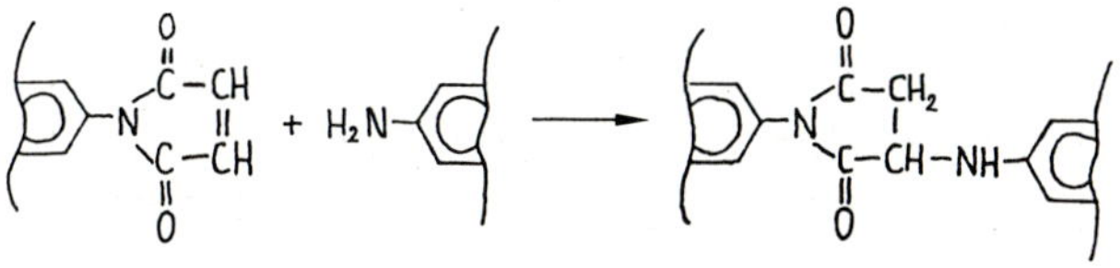

Scheme 12

High mechanical strength at dynamic stresses has the crosslinked material from BPA/DC, BMI, polyetherimide and a polyacrylate [66].

6 Epoxide Resin Compositions with Polyfunctional Cyanates

The reaction of epoxide resins with dicyanates was investigated using a model system. The system consisted of phenyl glycidyl ether and cumylphenyl cyanate [67]. Three reactions proceeded in the system:

(a) cyanate cyclotrimerization
(b) addition of epoxy and hydroxyl groups (ether formation)
(c) reaction of cyanate with epoxide resulting in oxazoline ring formation (Scheme 13)

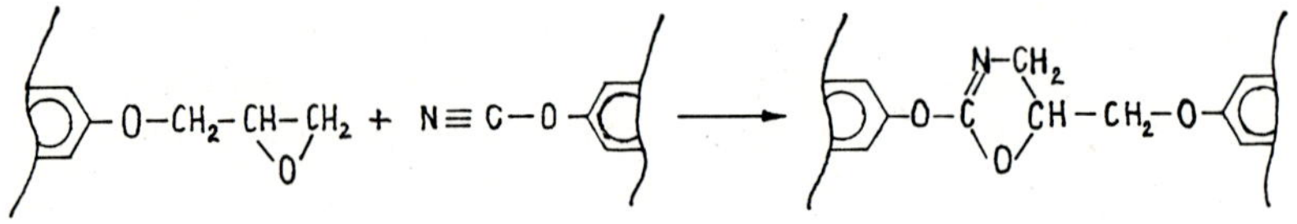

Scheme 13

Reaction (a) proceeded initially at a faster rate than reaction (c). The epoxy groups, which did not react with the fast disappearing cyanate, were consumed in the reaction (b). The fraction of epoxy groups consumed in the reaction (c) can be increased, if appropriate catalysts are added.

Full conversion can be achieved at high curing temperatures only. The BPA/DC — BPA/ECH epoxide resin ratio should be in the range of (25–45):(75–55).

Different reaction scheme was proposed by Bauer for the non-catalyzed dicyanate-epoxide system (Scheme 14) [68, 69]. That mechanism, which involves a phenol abstraction, should be considered with the Scheme 6 above concerning dicyanate polycyclotrimerization [12]. The mechanism proposed by Bauer concerns, however, the reaction up to the gelation only.

The statistical structural model for the gelation behavior of cyanate-epoxide polyreactions was described by Bauer [69] using the cascade formalism, according to the approach by Dušek [70], which had been previously applied for the polycyclotrimerization of acetylene derivatives.

An investigation of the thermal and thermooxidative stability of crosslinked polymers obtained from BPA/DC and BPA/DC compositions with BPA/ECH epoxide resin or epoxynovolak resin has demonstrated that the epoxy resin decreased the thermal stability of BPA/DC based non-modified polycyanurate [71]. The systems with epoxynovolak resin behave somehow better.

The authors claim that the thermal stability of polymers with aroxy-triazine groupments is much better than that with alkoxy-triazine fragments. It can be assumed that the proper way to increase the thermal stability is to enhance oxazoline ring formation.

The properties of the epoxide/cyanate compositions are inferior to those of non-modified cyanate polymers. They are, however, better in comparison with the properties of crosslinked BPA/ECH epoxide resin and epoxynovolaks [72]. Equimolar epoxide/cyanate ratio was recommended. Outstanding water resistance and dielectric properties at elevated temperatures are emphasized. In many patent applications, improved solvent resistance is mentioned.

The epoxide/cyanate compositions contain catalysts of cyanate cyclotrimerization and oxazoline ring formation. Curing agents for epoxide resin are mentioned in several patents. Some curing agents (e.g. pyromellitic dianhydride, cf. Sect. 3) react not only with epoxide resin, but also with dicyanate, thus contributing to the increase in the crosslinking density.

A composition consisting of low-molecular-weight BPA/ECH epoxide resin, tetraethylenepentaamine as a curing agent and 25% BPA/DC was suggested [73]. A similar composition contains tetraethylammonium bromide and 1-(2-cyanoethyl)-2-

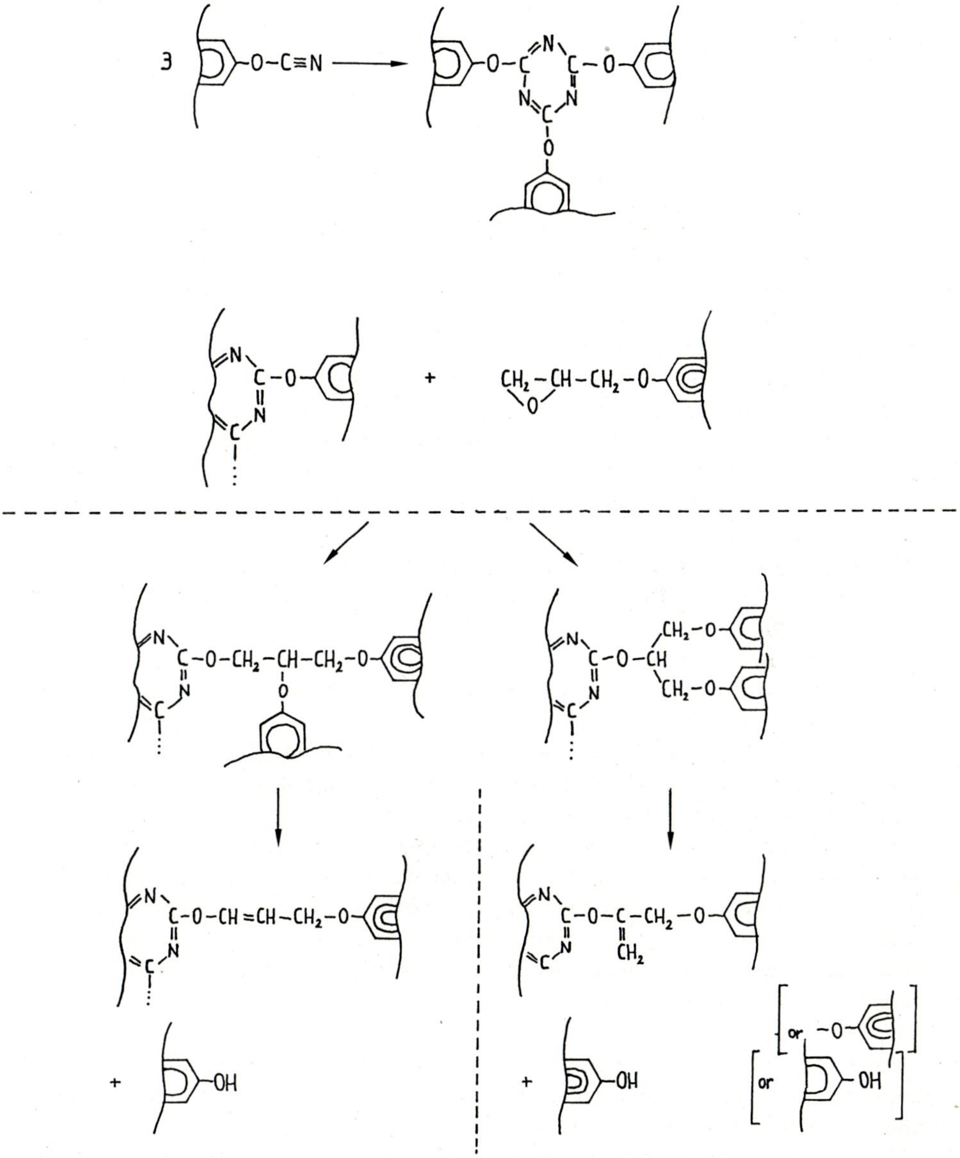

Scheme 14

phenylimidazole [74] or no curing agent [75]. A binder based on Epiclon 827, BPA/DC and Zn octoate was used for the manufacture of copper clad laminates [76]. To increase T_g, mixtures of polyfunctional epoxide resins, e.g. triglycidyl-p-aminophenol, i.e. p-[N,N-bis(2,3-epoxypropyl)aminophenyl] 2,3-epoxypropyl ether, and o-cresol based epoxynovolak resin, were used with a small amount of BPA/DC; difunctional epoxides: BPA diglycidyl ether or the cycloaliphatic 3,4-epoxycyclohexylmethyl 3,4-epoxycyclohexanecarboxylate were also added. Dicyandiamide was used as a curing agent for the epoxide resins [77].

It is advisable that the epoxide resin/hardener component has high T_g. A crosslinked IPN was obtained e.g. from epoxynovolak resin, pyromellitic dianhydride, triethanolamine (epoxide — acid anhydride reaction catalyst), BPA/DC and Zn acetate (cyclotrimerization catalyst) [78].

Curing agents and cyclotrimerization catalysts should be properly chosen for the epoxide-BPA/DC systems. As an example, a mixture consisting of Zn octoate, 2-ethylimidazole and triethylenediamine (DABCO) can be mentioned [79].

Monofunctional p-isopropenylphenyl cyanate was used with BPA/DC in a composition with BPA diglycidyl ether and Co naphthenate catalyst [80].

In several patents, the application of BPA/DC prepolymer in epoxide resin compositions is described [79, 81–84]. This approach is similar of that mentioned in the chapter devoted to cyanate/maleimide compositions [62].

As an additional component, various thermoplastic polymers can be used. As a binder for copper clad laminates, a solution of solid epoxide resin (Epikote 1001), BPA/DC prepolymer, Zn acetate and poly(phenylene sulfide) was used [83]. Other binders for reinforced plastics contain polysulfone. Such compositions consist of liquid BPA/ECH epoxide resin, BPA/DC prepolymer, polysulfone and bis(4-hydroxyphenyl)sulfone [85]. Bis(4-aminophenyl)sulfone can be also added [86]. In such systems the bisphenol reacts with the epoxide resin as a chain extension agent, whereas the diamine crosslinks the diepoxide. The T_g values are close to 200 °C. They can be increased a little, if the BPA/ECH epoxide resin is replaced by the tetraepoxide: N,N,N',N'-tetrakis(2,3-epoxypropyl)diaminodiphenylmethane [87].

Saturated polyesters can be also added to the epoxide/cyanate prepolymer systems [81, 88].

Furthermore, flexibilized compositions have been described. As flexibilizers, carboxyl terminated polybutadiene [89] or simply polybutadiene [79] were mentioned.

The cyanate and the epoxide network can be bound by the reaction of phenolic hydroxyls contained in the partially cyclotrimerized cyanate prepolymer with the epoxy groups. In the first stage, a prepolymer was obtained by heating a mixture of BPA monocyanate and dicyanate with Co naphthenate as a catalyst (Scheme 15).

In the second stage, the above prepolymer is prereacted with liquid epoxide resin in the presence of benzyltrimethylammonium chloride as a catalyst. The crosslinking occurs as a result of the reaction of the remained epoxy groups with 4,4'-diaminodiphenylmethane added [90].

Fire resistant reinforced plastic was obtained by using tetrabromo-BPA based epoxide resin, BPA/DC monomer and prepolymer as well as an imidazole curing agent and Zn octoate [82].

As mentioned above, the main application of the epoxide/dicyanate systems are copper clad laminates. Other important uses are conductive materials with silver

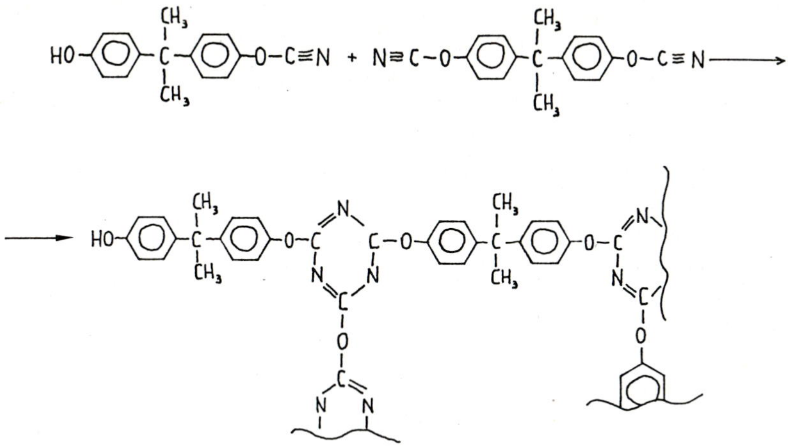

Scheme 15

powder [84, 88] and impregnation of porous polytetrafluoroethylene for copper or aluminum clad laminates [91].

7 Three-Component Cyanate/Epoxide/Maleimide Systems

In addition to the two-component cyanate/epoxide, cyanate/maleimide and epoxide/maleimide systems, numerous three-component compositions have been described in 34 Japanese and 2 FRG patent applications. The patent applications differ in curing agents, other additives and applications of the binders obtained. Sometimes the compositions mentioned in Sect. 4 above are given here again, differing in that a bismaleimide is added.

The three-component compositions can be considered as having been derived from two-component epoxide/maleimide systems with BPA/DC as the third component. There are two different epoxide/maleimide systems known. In the first one, BMI is reacted with a molar excess of an aromatic diamine. Thus, oligoaspartimide diamine is formed (cf. Scheme 12). The diamine serves then as a crosslinking agent for the epoxide resin [92]. Another approach consists in reacting of excess BMI with the aromatic diamine. In such case, oligoaspartimide with terminal maleimide groupments is obtained. In the second stage, reaction of residual NH groups in oligoaspartimide with the epoxide resin added and the polymerization of terminal maleimide groups occurs.

There have no results of investigations of the cyanate/epoxide/maleimide co-curing mechanism been published. It can be assumed that the following reactions may take place (cf. Sects. 5 and 6):
— bismaleimide polymerization through the double bonds,
— dicyanate cyclotrimerization,
— maleimide/cyanate cycloaddition (Scheme 10),
— epoxide/cyanate cycloaddition with oxazoline ring formation (Scheme 13)

— epoxide resin crosslinking (if curing agents are added), with possible addition of amine curing agent to maleimide (aspartimide formation, Scheme 12) and to cyanate (isourea formation, Scheme 8) or addition of acid anhydride curing agent to cyanate (Scheme 7).

A typical three-component composition consists of BPA/DC prepolymer, bis(4-maleimidophenyl)methane (BMI), epoxide resin, Zn acetate and triethylenediamine in methylethylketone [93].

A binder was obtained from an epoxynovolak resin, BPA/DC, BMI, dicyclopentadiene, Zn acetate and dicumyl peroxide [94]. In a similar composition, BPA/DC prepolymer was used; as curing and cyclotrimerization initiators and catalysts, catechol, triethylenediamine, Zn acetate and benzoyl peroxide are mentioned [95]. Other compositions contain Al acetylacetonate and a silicone resin [96], p-toluenesulfonic acid monohydrate and Zn octoate (for rapid curing) [97], or dicumyl peroxide and Zn octoate [98].

Polyethylene and a peroxide initiator were added to the usual BPA/DC-BMI-epoxide resin composition. The use of unsaturated polyesters from tetrabromophthalic anhydride, maleic anhydride and ethylene glycol as well as of other fire retardants was also described [99].

A binder for copper clad laminates contains the prepolymer from BPA/DC and N-(3,5-dimethyl-4-vinylphenyl)maleimide in methylethylketone, an epoxynovolak resin, Zn acetate and tert.butyl peroxide [100]. In a similar composition, a prepolymer obtained from epoxide resin, BMI and bis(4-aminophenyl)methane was mixed with BPA/DC, Zn acetate and tert.butyl peroxide in solution [101].

The heat resistance of a BPA/DC-BMI-epoxynovolak composition can be considerably increased by adding pyromellitic dianhydride [102].

Fire resistant polymers were obtained from brominated epoxynovolak resin, BPA/DC prepolymer, BMI, Zn acetate and benzoyl peroxide [103] or from an oligoaspartimide (BMI-diamine reaction product), BPA/DC, 2,2-bis(3,5-dibromo-4-hydroxyphenyl)propane (i.e. Tetrabromo-Bisphenol A) and 2-ethyl-4-methylimidazole [104]. A mixture of BPA/DC, BMI and epoxide resin with brominated polycarbonate, copoly[oxy-2,6-dimethylphenylene)-(oxy-2,3,6-trimethylphenylene)] and a catalyst was also suggested [105].

Epoxynovolak resin and BPA/DC-BMI prepolymer, tert.butyl peroxide and Zn acetate [106, 107] or 2-phenylimidazole and other catalysts [108] were filled with wollastonite. Carbon-fiber reinforced composites were obtained using a binder, which consisted of BPA/DC, BMI, an epoxynovolak, 2-ethyl-4-methylimidazole and an organic solvent [109]. A BPA/DC-BMI prepolymer in methylethylketone was mixed with middle-molecular-weight epoxide resin (Epikote 1001), 2-ethyl-4-methylimidazole, Zn acetate and triethylenediamine; thermal shock resistant GRP was thus obtained [110].

As in the cyanate/epoxide compositions, polysulfone is used as an additive in cyanate/epoxide/maleimide systems. As an example, a composition for carbon-fiber composites should be mentioned. It contains BPA/DC-BMI prepolymer, epoxide resin and polyethersulfone [111–113]; Zn acetate and dicumyl peroxide are added. Polyethersulfone powder was added to the three-component system with 4,4'-diaminodiphenylmethane-based tetraepoxide as one of the epoxy resins used [114].

Very high T_g (288 °C) shows the polymer obtained from BPA/DC-BMI prepolymer, epoxide resin, bis(4-aminophenyl)sulfone (4,4'-diaminodiphenylsulfone, DDS) and polyethersulfonediol diglycidyl ether (reaction product of phenolic hydroxy-terminated polyethersulfone with epichlorohydrin in alkaline medium) [115].

Molar excess of epoxynovolak resin was reacted with DDS; highly branched aminoepoxide resin was thus obtained. The resin was then mixed with a non-modified epoxide resin, BPA/DC-BMI copolymer, benzoyl peroxide and N-(3,4-dichloro-phenyl)-N',N'-dimethylurea (curing agent for the epoxy component) [116].

Another multicomponent heat resistant (T_g = 220 °C) system consists of epoxide resin, carboxyl terminated polybutadiene, glycidyl methacrylate, BPA/DC, diamines and bismaleimides. It contains catalysts and initiators, e.g. Zn octoate, triethylene-diamine and benzoyl peroxide [89].

Impact resistance of carbon-fiber composites or GRP with the three-component polymer matrix can be improved by the incorporation of a terephthalic acid/neo-pentyl glycol polyester [117]. The polyester was carboxyl terminated by the addition of trimellitic anhydride to the terminal hydroxyls [118].

Two-layered GRPs for copper clad laminates are obtained with one layer consisting of the three-component system (e.g. BPA/DC, BMI, brominated epoxide resin, Zn octoate and triethylenediamine in methylethylketone). The other layer has the usual epoxy matrix (brominated epoxide resin, dicyandiamide as a hardener and 2-ethyl-4-methylimidazole as curing accelerator) [119]. As similar two-layered laminate contains BPA/DC, BMI, epoxynovolak resin, Zn acetate and triethylenediamine in the first layer and BPA/DC only with the same catalysts in the second layer [120].

A coating composition consists of BPA/DC-BMI prepolymer, epoxide resin and Zn acetate in dimethylformamide; a solution of polyhydantoin in CH_2Cl_2 is added [121].

The three-component cyanate/maleimide/epoxide compositions are mainly used as polymer matrix in copper clad laminates and in carbon fiber composites for engineering purposes. High heat resistance, water and solvent resistance, mechanical and impact strength is claimed. A composition for copper wire enamelling [121] and a resin for electric motor coil windings impregnation were described [107].

8 Polyfunctional Unsaturated Monomers and Cyanates

Semi-IPNs are obtained by simultaneous polymerization of styrene and polycyclo-trimerization of BPA/DC (Sect. 2) [52, 53]. IPNs consisting of two separate networks are formed, if unsaturated monomers with two or more polymerizable double bonds are used.

Heat resistant IPN systems were obtained by simultaneous radical polymerization of divinylbenzene with benzoyl peroxide as an initiator and Zn acetate as cyclotri-merization catalyst [122]. Hot-curing composition contains BPA/DC, BMI, epoxide resin, Zn acetate and divinylbenzene [123]. Crosslinked compositions consisting of BPA/DC and BPA bis(vinylbenzyl) ether show T_g values above 240 °C [124].

High crosslinking density was reached using BPA/DC, trimethylolpropane tri-methacrylate and p-toluenesulfonic acid monohydrate [125]. A composition based on BPA/DC and 2,2-bis[4-(2-hydroxyethoxy)phenyl] propane dimethacrylate with a peroxide initiator and Fe acetylacetonate catalyst contains, moreover, a block copolymer of butadiene and α-methylstyrene as well as PPO. The composition is suggested to be used in trichloroethylene solution as a binder for copper clad laminates [126].

Curing of multicomponent systems consisting of BPA/DC, BMI and diallyl or triallyl esters (triallyl isocyanurate TAIC in particular) results in high T_g values. Such compositions consist of BPA/DC, BMI, TAIC, diallyl phthalate and dicumyl peroxide [127], BPA/DC prepolymer, BMI, TAIC, *tert*.butyl peroxide, Zn acetate and DABCO [128] or BMI/trimethylene bis(4-aminobenzoate) prepolymer, BPA/DC, Zn octoate and TAIC [129].

As the unsaturated monomers in the BPA/DC systems, allyl ethers of bisphenols, e.g. bis(4-allyloxyphenyl)sulfone, were used. $T_g = 260$ °C was reached [130].

Cold curing unsaturated resin was obtained in the following way. A mixture of BPA, BPA monocyanate and BPA dicyanate was cyclotrimerized and reacted with BMI. The obtained prepolymer with phenolic hydroxyls (cf. Scheme 9) was then treated with epichlorohydrin and alkali. The phenolic hydroxyls were thus transformed into glycidyl ether groups. Then the addition of methacrylic acid to the epoxy groups was carried out. The obtained vinyl ester (epoxyacrylate) type resin was dissolved in styrene and cured with the usual benzoyl peroxide/dimethylaniline system [131].

Simultaneous IPN was obtained by dissolving BPA/DC in an unsaturated polyester resin (UPR) and crosslinking at elevated temperature using benzoyl peroxide and a zinc compound as cyclotrimerization catalyst. Prolonged post-curing at 180 °C was needed to reach elevated T_g values [132, 133].

It can be assumed that two separate networks with no covalent bonds between the UPR and the cyanate-based triazine network are formed. The possible addition of terminal hydroxyls from the unsaturated polyester to the —C≡N bonds in BPA/DC is rather improbable as the addition of alcohols to cyanates, leading to iminocarbonate derivatives (Scheme 8), only occurs in the presence of strong alkali catalysts [134]. The cyanate cyclotrimerization has been evidenced from disappearance of the 2230 and 2270 cm^{-1} and the appearance of 1370 and 1560 cm^{-1} absorption bands in the infrared spectra of the crosslinked IPN.

The torsional measurements show only one maximum on the damping curve. This means that there is a good segmental miscibility between the networks in the UPR-BPA/DC system.

Thermomechanical, torsional and derivatographic measurements were carried out. The addition of BPA/DC results in an increase in T_g values by 32 °C at 40 phr dicyanate and by 54 °C at 80 phr dicyanate. The disappearance of the maximum in the high-elastic region of the thermomechanical curve during the repeated measurement confirms that the high-temperature postcuring is needed to reach complete crosslinking.

The comparison of the results of derivatographic measurements shows that the characteristic temperature values (beginning of weight loss at 115–128 °C, sharp decrease in weight loss rate at 170–190 °C, beginning of fast decomposition at 270–278 °C and maximum decomposition rate at 355–358 °C) are only little changed as a result of the BPA/DC addition. Weight loss of the BPA/DC-containing systems up to the temperature of beginning of fast decomposition is significantly decreased in consequence of postcuring.

The addition of 40 phr BPA/DC improves most mechanical parameters of the UPR;

80 phr BPA/DC does not give much better values. Tensile strength increases by 35%, whereas the increase in modulus of elasticity as well as bending and compression strength is not as significant. Decrease in elongation at break and increase in impact strength was observed [135].

An improvement of some dielectric properties was found. BPA/DC (40 phr) shifts the low tg δ range towards higher temperatures. Sharp decrease in volume resistivity occurs at higher temperatures in comparison with the same UPR containing no BPA/DC.

9 Compositions of Polyesters, Polyurethanes, Modified Phenolic Resins and Related Materials with Dicyanates

Branched saturated polyester from adipic acid, phthalic anhydride, neopentyl glycol and trimethylolpropane was mixed with BPA/DC solution in 2-methoxyethanol/xylene and diluted with ethyl acetoacetate. A thermohardening coating with high hardness and good adhesion on steel was obtained [136]. A similar coating composition consists of BPA/DC and a polyester from adipic and isophthalic acid, trimellitic anhydride, neopentyl glycol and trimethylolpropane [137, 138].

A flexibilized thermohardening composition contained ca. 90% BPA/DC and a polyester from dimerized fatty acids and hexanediol. Elongation at break and impact strength increased considerably, whereas the heat deflection temperature was only little decreased [139].

An electroinsulating coating was obtained by mixing a polyimide resin with BPA/DC or BPA/DC-BMI prepolymer. The low-molecular-weight polyimide resin contained reactive end groups ($-C\equiv CH$, $-COOH$, $-NH_2$ etc.) [140].

Polyurethane-polytriazine was synthesized from polyoxypropylene diol, 4,4'-diisocyanatodiphenylmethane (MDI), 1,4-butanediol and BPA/DC [141].

Binders for copper clad laminates consist of a solution of phenolic resol (part A) and of a composition obtained from a drying oil modified resol, hexamethylenetetraamine and BPA/DC monomer or prepolymer [142, 143].

10 Rubbers Modified with Dicyanates

Some dicyanate-containing compositions, which contain rubbers as flexibilizing components, were described in the preceding chapters. There were also patent applications made, where dicyanates were claimed as additives in typical rubber mixtures. In such mixtures, butadiene-acrylonitrile rubber is used. The main components of such binders are nitrile rubber, BPA/DC and methylethylketone. They contain, moreover, Zn octoate and Fe_2O_3 [144] or ZnO and sulfur [145]. Isoprene-acrylonitrile rubber, BPA/DC prepolymer, Zn octoate, DABCO and benzoyl peroxide were dissolved in a methylethylketone-dimethylformamide mixture. Glass fiber was impregnated with the obtained solution [146].

A rubber mixture contains ethylene-propylene rubber, BPA/DC, $CaCO_3$, ZnO, sulfur, dicumyl peroxide and mineral oil. The final product has elevated thermal stability [147].

11 References

1. Vinogradova SV et al. (1971) Izv. Akad. Nauk SSSR, Ser. Khim: 837
2. Martin D, Bauer M (1983) Org. Syntheses 61: 35
3. Pankratov VA, Vinogradova SV, Korshak VV (1977) Usp. Khim. 46: 530
4. Korshak VV et al. (1978) J. Polym. Sci., Polym. Chem. Ed. 16: 1697
5. Cozzens RF, Walter P, Snow AW (1987) J. Appl. Polym. Sci. 34: 601
6. Korshak VV et al. (1972) Dokl. Akad. Nauk SSSR 202: 347
7. Bonetskaya AK et al. (1973) Dokl. Akad. Nauk SSSR 212: 1353
8. Bauer M, Bauer J, Kühn G (1986) Acta Polymerica 37: 218
9. Frenkel SM et al. (1976) J. Prakt. Chem. 318: 923
10. Lebedev BV et al. (1977) Dokl. Akad. Nauk SSSR 237: 383
11. Bauer M, Bauer J, Garske B (1986) Acta Polymerica 37: 604
12. Bauer M, Bauer J, Kühn G (1986) Acta Polymerica 37: 715
13. Bauer M, Bauer J, Much H (1986) Acta Polymerica 37: 221
14. Korshak VV et al. (1975) Vysokomol. Soed. A17: 482
15. Korshak VV et al. (1974) Vysokomol. Soed. A16: 15
16. Grigat E, Pütter R (1964) Chem. Ber. 97: 3012
17. Grigat E, Pütter R (1967) Angew. Chem. 79: 219
18. Kubens R et al. (1968) Kunststoffe 58: 827
19. US Pat. 4608434
20. Jpn. Pat. Appl. 8810658; Ch. Abs. 108: 222610
21. US Pat. 4709608
22. Jpn. Pat. Appl. 81127630; Ch. Abs. 96: 36283
23. Jpn. Pat. Appl. 82200451; Ch. Abs. 98: 216559
24. Brand RA, Harrison ES (1982) Sci. Tech. Aero-Space Rep.; Ch.Abs. 98: 180312
25. Ger. Pat. Appl. 3613006
26. Grigat E (1970) Angew. Chem. 82: 81
27. Korshak VV et al. (1973) Vysokomol. Soed. B15: 319
28. Pankratov VA (1975) Vysokomol. Soed. A17: 2189
29. US Pat. 3491060
30. US Pat. 3502617
31. Ger. Pat. 1220132
32. USSR Pat. 520375
33. Grigat E, Pütter R (1964) Chem. Ber. 97: 3018
34. Sperling LH (1981) Interpenetrating polymer networks and related materials, Plenum, New York
35. Lipatov YuS, Sergeeva LM (1979) Vzaimopronikayushchie Polimernye Setki. Naukova Dumka, Kiev
36. Klempner D (1978) Angew. Chem. 90: 104
37. Penczek P (1981) Polimery (Warsaw) 26: 157
38. Hsiueh HS, Miller RL, Segal L (1985) Int. Conf. Compos. Mat., Conf. Proc., 5th 1655; Ch. Abs. 104: 51663
39. Jpn. Pat. Appl. 86183993; Ch. Abs. 106: 19675
40. Cercens JL, Huang SJ (1984) Polym. Prepr. 25: 114
41. Ger. Pat. Appl. 3344313
42. Ger. Pat. Appl. 3608253
43. Jpn. Pat. Appl. 8358627; Ch. Abs. 99: 196224
44. US Pat. 4110394
45. Jpn. Pat. Appl. 81145945; Ch. Abs. 96: 69894
46. Jpn. Pat. Appl. 8367751; Ch. Abs. 99: 213493
47. Jpn. Pat. Appl. 8525744; Ch. Abs. 102: 186363
48. Jpn. Pat. Appl. 83164638; Ch. Abs. 100: 192968
49. Jpn. Pat. Appl. 8316646; Ch. Abs. 100: 193168
50. Eur. Pat. Appl. 266986
51. GDR Pat. 239601

52. Ger. Pat. Appl. 2628417
53. US Pat. 4581425
54. Ger. Pat. Appl. 3412907
55. Ger. Pat. Appl. 3426883
56. US Pat. 4559399
57. Ayano S (1985) Kunststoffe 75: 475
58. Jpn. Pat. Appl. 8198162; Ch. Abs. 95: 220998
59. Jpn. Pat. Appl. 82165451; Ch. Abs. 98: 144393
60. Jpn. Pat. Appl. 8433359; Ch. Abs. 101: 92132
61. Jpn. Pat. Appl. 8376453; Ch. Abs. 100: 7776
62. Eur. Pat. Appl. 56473 (1982)
63. Jpn. Pat. Appl. 82202343; Ch. Abs. 98: 216552
64. Jpn. Pat. Appl. 86229944; Ch. Abs. 106: 120986
65. Jpn. Pat. Appl. 8532831; Ch. Abs. 103: 38158
66. Eur. Pat. Appl. 266986 (1988)
67. Shimp DA, Hudock FA, Ising SJ (1988) Int. SAMPE Symp. Exhib. 33: 754; Ch. Abs. 110: 96323
68. Bauer M (to be published)
69. Bauer J, Bauer B (1988) Acta Polymerica 39: 548
70. Dušek K et al. (1977) Vysokomol. Soed. A19: 1368
71. Bauer M, Gnauck R (1987) Acta Polymerica 38: 658
72. Bauer M (1987) Acta Polymerica 38: 393
73. US Pat: 4142034
74. Jpn. Pat. Appl. 80112256; Ch. Abs. 94: 31585
75. Jpn. Pat. Appl. 8526031; Ch. Abs. 103: 23324
76. Jpn. Pat. Appl. 7639770; Ch. Abs. 85: 47799
77. Ger. Pat. 2205431
78. Jpn. Pat. Appl. 81112923; Ch. Abs. 95:220860
79. Jpn. Pat. Appl. 81106918; Ch. Abs. 95: 204907
80. US Pat. 4558115
81. Ger. Pat. Appl. 3509220
82. Jpn. Pat. Appl. 75116541; Ch. Abs. 84: 60535
83. Jpn. Pat. Appl. 85125661; Ch. Abs. 103: 197100
84. Ger. Pat. Appl. 3423385
85. Jpn. Pat. Appl. 8868623; Ch. Abs. 109: 171586
86. Jpn. Pat. Appl. 8854418; Ch. Abs. 109: 171775
87. Jpn. Pat. Appl. 87277466; Ch. Abs. 109: 38910
88. Jpn. Pat. Appl. 8869883; Ch. Abs. 109: 191741
89. Jpn. Pat. Appl. 81141310; Ch. Abs. 96: 53279
90. US Pat. 4506063
91. Jpn. Pat. Appl. 81141310; Ch. Abs. 96: 53279
92. Penczek P, Kosińska W, Drągowska E (1974) Polimery (Warsaw) 19: 613
93. Jpn. Pat. Appl. 81110991; Ch. Abs. 95: 8424
94. Jpn. Pat. Appl. 8374750; Ch. Abs. 99: 196248
95. Jpn. Pat. Appl. 8184727; Ch. Abs. 95: 170643
96. Jpn. Pat. Appl. 82133119; Ch. Abs. 98: 55084
97. Jpn. Pat. Appl. 82202342; Ch. Abs. 98: 216553
98. Jpn. Pat. Appl. 8736460; Ch. Abs. 107: 177518
99. Ger. Pat. Appl. 2952440
100. Ger. Pat. Appl. 3517395
101. Jpn. Pat. Appl. 85184524; Ch. Abs. 104: 130851
102. Jpn. Pat. Appl. 8195951; Ch. Abs. 95: 204957
103. Jpn. Pat. Appl. 8854419; Ch. Abs. 109: 191485
104. Jpn. Pat. Appl. 75132099; Ch. Abs. 84: 60484
105. Jpn. Pat. Appl. 8607331; Ch. Abs. 105: 44023
106. Jpn. Pat. Appl. 85184523; Ch. Abs. 104: 69737
107. Jpn. Pat. Appl. 86278113; Ch. Abs. 107: 135431
108. Jpn. Pat. Appl. 85238322; Ch. Abs. 104: 150178

109. Jpn. Pat. Appl. 85 15439; Ch. Abs. 103: 23471
110. Jpn. Pat. Appl. 81 45848; Ch. Abs. 96: 20834
111. Jpn. Pat. Appl. 88 15846; Ch. Abs. 109: 161508
112. Jpn. Pat. Appl. 87257123; Ch. Abs. 108: 151850
113. Jpn. Pat. Appl. 88205328; Ch. Abs. 110: 40250
114. Jpn. Pat. Appl. 87185720; Ch. Abs. 108: 95571
115. Jpn. Pat. Appl. 85250026; Ch. Abs. 104: 225821
116. Jpn. Pat. Appl. 87172013; Ch. Abs. 108: 7030
117. Jpn. Pat. Appl. 87146927; Ch. Abs. 108: 57129
118. Jpn. Pat. Appl. 87146928; Ch. Abs. 108: 7013
119. Jpn. Pat. Appl. 81 77492; Ch. Abs. 95: 134406
120. Jpn. Pat. Appl. 81 84726; Ch. Abs. 95: 170644
121. Jpn. Pat. Appl. 81 125449; Ch. Abs. 96: 21424
122. Jpn. Pat. Appl. 81 86934; Ch. Abs. 95: 205555
123. Jpn. Pat. Appl. 83 02332; Ch. Abs. 99: 23512
124. PCT Int. Appl. WO 8807562; Ch. Abs. 110: 174379
125. Jpn. Pat. Appl. 83 02346; Ch. Abs. 99: 39303
126. Jpn. Pat. Appl. 88 83152; Ch. Abs. 109: 171608
127. Jpn. Pat. Appl. 85 113406; Ch. Abs. 104: 35115
128. Jpn. Pat. Appl. 85240741; Ch. Abs. 104: 131134
129. Ger. Pat. Appl. 3 507609
130. Eur. Pat. Appl. 269392 (1988); Ch. Abs. 110: 9159
131. US Pat. 4 661 553
132. Penczek P (1989) Adv. Interpenetr. Polym. Networks 1: 239
133. Polish Pat. Appl. P-273447 (1988)
134. Grigat E, Pütter R (1964) Chem. Ber. 97: 3018
135. Penczek P, Mirkowska B (in preparation)
136. Jpn. Pat. Appl. 78 111 334; Ch. Abs. 90: 56471
137. Jpn. Pat. Appl. 80 48258; Ch. Abs. 93: 73984
138. Jpn. Pat. Appl. 78 111 333; Ch. Abs. 90: 56472
139. Ger. Pat. Appl. 3 413547
140. Ger. Pat. Appl. 2 945797
141. Ger. Pat. Appl. 2 620487
142. Jpn. Pat. Appl. 75 173050; Ch. Abs. 84: 123058
143. Jpn. Pat. Appl. 75 134051; Ch. Abs. 84: 136869
144. Jpn. Pat. Appl. 85 233130; Ch. Abs. 104: 188329
145. US Pat. 4 649 714
146. Ger. Pat. Appl. 3 628 362
147. Jpn. Pat. Appl. 82 170748; Ch. Abs. 98: 161998

Editor: K. Dušek
Received November 6, 1989

Polymers Containing Metallochelate Units

A. D. Pomogailo[1] and I. E. Uflyand[2]

[1] Institute of Chemical Physics, USSR Academy of Sciences, 142432 Chernogolovka, Moscow Region, USSR
[2] Department of Chemistry, State Pedagogical Institute, Engelsa 33, 344082 Rostov-on-Don, USSR

Major advances and problems in the field of synthesis, properties, structure and applications of polymers containing metallochelate units are discussed. Included are terminology, classification and nomenclature of these compounds as well as major approaches to calculating the equilibrium constants of chelation with polymeric ligands and chelate effect in metallopolymeric systems. Special attention is paid to the production and structural features of polymers containing metallochelate units. The most important applications of such polymers are classified.

1 Introduction 63
2 General Description of Polymers Containing Metallochelate Units . . . 63
   2.1 Terminology, Classification, Nomenclature 63
   2.2 Main Approaches to Calculating Equilibrium Constants
       in Metal Ion — Chelating Macroligand Systems 65
   2.3 Chelate Effect in Metallopolymeric Systems 68
3 Preparation of Polymers Containing Metallochelate Units 77
   3.1 Syntheses Using Chelating Macroligands 77
   3.2 Polymer-Analogous Reactions 82
   3.3 Methods of Assembly 83
   3.4 Synthesis of Polynuclear Metallochelate Polymers 90
4 Structure of Polymers Containing Metallochelate Units 93
   4.1 Structural Features of Metallochelate Containing Soluble
       Polymers 93
   4.2 Details of Metal Binding to Cross-Linked Chelating
       Polymers 95
5 Trends in the Use of Polymers Containing Metallochelate Units . . . 97
   5.1 Upgrading Engineering and Physico-Mechanical Polymer Properties 98
   5.2 Catalytic Activity of Macromolecular Metallochelates 99
   5.3 Other Applications 99
6 Conclusion 100
7 References 101

List of Abbreviations and Symbols

AA	— acrylic acid
acacH	— acetylacetone
AN	— acrylonitrile
CMCS	— chloromethylated copolymer of styrene and divinylbenzene
CML	— chelating macroligand
—co—	— copolymer
DH_2	— dimethylglyoxime
Dipy	— 2,2'-dipyridyl
DMF	— dimethylformamide
DVB	— divinylbenzene
en	— ethylenediamine
—g—	— —graft—
$\Delta H, \Delta G, \Delta S$	— change of enthalpy, free energy, and entropy, respectively
$\bar{K}, (K_{eq})$	— constant of stability (equilibrium)
L	— ligand or functional group of polymer
l	— chain length
LMWM	— low-molecular weight metallochelate
M	— metal
MCM	— metal chelate monomer
MMA	— methylmethacrylate
MX_n	— metal compound
$\tilde{n}$	— formation function
n	— metal valency
℗, ~	— polymeric chain
PCMSt	— polychloromethylstyrene
PCMU	— polymers containing metallochelate units
PDA	— poly (diethyl azelate)
PE	— polyethylene
PHPG	— poly-N^5-(3-hydroxypropyl)-L-glutamine
PIP	— pyrrolylmethyliminopropene
PIS	— pyrrolylmethyliminostyrene
PMA	— poly (methyl vinyl ketone)
PVAK	— poly (vinyl acetonyl ketone)
Py	— pyridine
St	— styrene
St—co—DVB	— copolymer of styrene and divinylbenzene
St—co—4-VP	— copolymer of styrene and 4-vinylpyridine
T_g	— glass transition point
THF	— tetrahydrofuran
VIm	— vinylimidazole
4-VP (P4VP)	— 4-vinylpyridine (poly-4-vinylpyridine)
VPd	— vinylpyrolidone
▨	— polyethylene surface
Θ	— conversion

1 Introduction

In the last two decades the chemistry of polymers containing metallochelate units (PCMU) has been receiving widespread attention [1]. Owing to the efforts of specialists in various fields such as polymer, coordination, bioinorganic and analytical chemistry, catalysis, etc., considerable advances have been achieved. So far a comparatively large number of PCMU containing practically all elements of the Periodic Table have been synthesized, quantitative characteristics of the chelation processes involving polymeric ligands obtained and their major structural features disclosed. PCMU turned out to be convenient for studies of a number of theoretical problems related to the polymer, coordination, bioinorganic, analytical chemistry, photochemistry, catalysis and other fields. Among such problems are macromolecule reactivity, competing coordination, analytical reagent selectivity, coordination compound stereochemistry, preparation of polyfunctional catalysts, etc. Moreover, these compounds can be regarded as models of substances participating in some biochemical processes.

Now it is difficult to conceive any field of human activity where PCMU are not used. Among them are health care and pharmacology (administering of desired metals into the organism, such as iron in the case of asiderotic anemia, as well as removal of harmful, particularly radioactive metals); atomic industry and hydrometallurgy (concentration and separation of rate metals and radioactive isotopes), chemistry and catalysis (preparation of highly effective immobilized catalysts) etc.

Studies of PCMU were started during World War Two, primarily under the pressure of practical needs. Thus the U.S. Air Force gave an order to search for polymers capable of withstanding temperatures as high as 600 °C [2]. Thereby studies in the field of PCMU were obviously encouraged although no final solution to the problem was found at that time. The first publications describing the results of these studies appeared in the end of the 1950s [3]. Approximately the same period was marked by the initiation of studies into chelating resins and the metal complexes based on them in the general context of the analytical problems such as concentration and separation of metals [4]. Later the PCMU chemistry was mainly stimulated by various industries. Thus in the early 1970s extensive studies of the catalytic properties of PCMU in a great variety of chemical reactions were carried out [5]. Large-scale studies of photochemical reactions and solar energy storage with the use of PCMU were started in the beginning of the 1980s [6].

This review is concerned with the major advances in preparation, investigation of properties, structure and applications of PCMU and future developments in this field.

2 General Description of Polymers Containing Metallochelate Units

2.1 Terminology, Classification, Nomenclature

PCMU are those high-molecular weight compounds that incorporate metallochelate cycles. The polymer chain in PCMU make them behave in many cases as ordinary high-molecular weight compounds, and the presence of metal ion in the chain is responsible for a number of properties typical of a given metal ion. However, such a combination introduces some specific features such as those related to a biocidic and catalytic activity, etc.

Depending on the position of metal with respect to the main chain, PCMU can be subdivided into two distinct classes. Polymeric chelates whose main chain contains a metal and which breaks upon its removal are termed coordination polymers. They have been exhaustively described elsewhere [7]. The second class of PCMU which is discussed in this review contains metal in a side chain. In this case the metal can be fairly readily removed or displaced by other metals so that the main chain remains intact. Such PCMU actually incorporate complexes of metals with chelating ion-exchange resins [8].

PCMU can be classified by the same scheme which is suitable for low-molecular weight metallochelates (LMWM) [9]. Therefore they are subdivided into molecular, intra-complex, macrocyclic and polynuclear types which in turn are grouped depending on the nature of the donor atoms (O,O-; N,N-; N,O-; P,P-chelates, etc.).

The molecular-type PCMU include compounds where the metal ion is linked to the donor atoms of the chelating fragment via the coordination bonds only. Such complexes are formed with the aid of fragments having the metallocyclization arrangement of the donor sites (**1**) or conformationally nonrigid groups with two or more donor sites (**2**).

Among the most comprehensively studied PCMU of this type are metal complexes with polymers containing ethylenediamine (en) [10], 2,2'-dipyridyl (Dipy) [11], dipyridylmethane [12] and other groups.

The intra-complex PCMU are those compounds in which at least one fragment is linked with the metal ion via both a valent and coordination bond. It should be noted that such compounds can be neutral, cationic, anionic or contain, in addition to chelating ligands, monofunctional varieties. The number of such PCMU is enormous therefore we shall indicate the most frequent types. A detailed consideration has been given earlier to complexes of metals with polymers containing diketone [13], o-hydroxyazomethine [14], enaminoketone [15], amino acid [16], 8-hydroxyquinoline [17] and other groups.

The macrocyclic-type PCMU give mostly polymers with the corresponding groupings. Phthalocyanines [18] and porphyrins [19] are the classical examples of ligands used to obtain macrocyclic complexes. Much attention is being paid to complexes incorporating cyclic polyethers [20] as well as their nitrogenous, phosphorus and sulfur analogues [21].

In the polynuclear-type PCMU each chelating fragment is linked with two or more metal ions, at least one of them being expected to form a metallocycle. It was not until recently that such compounds have received attention, and only limited studies have been made up to the present [15, 22].

The nomenclature of PCMU is based on that developed for complex compounds [23] with a prefix "poly" being added to the name of the corresponding substance. Thus polymers **3** and **4** are termed poly copper(II) p-vinylbenzoylacetonate [24] and poly palladium(II) dioximate [25], respectively:

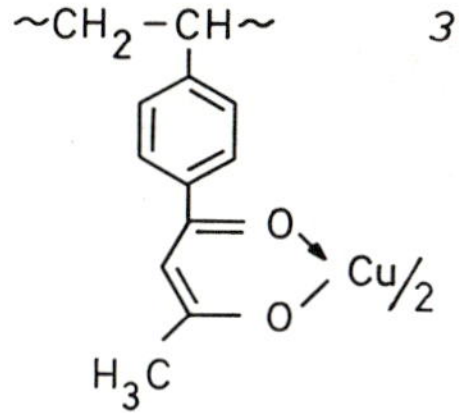

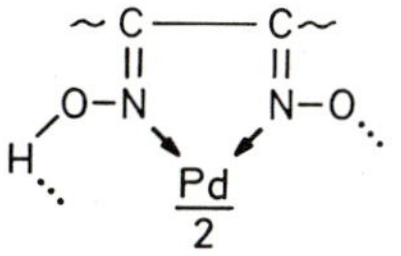

However, more frequently the name of a PCMU is made up of that of a polymer and metal compound (MX_n). Thus compounds **5** and **6** are termed a complex of iron(III) chloride with a dipyridyl methane-based polymer [12] and that of copper(II) with the condensation product of p-hydroxybenzoic acid, urea and formaldehyde [26], respectively:

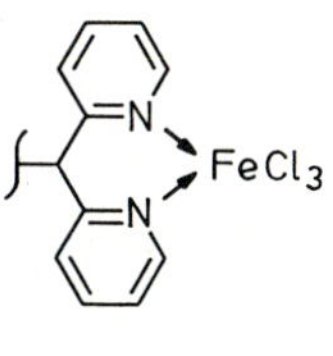

2.2 Main Approaches to Calculating the Equilibrium Constants in Metal Ion — Chelating Macroligand Systems

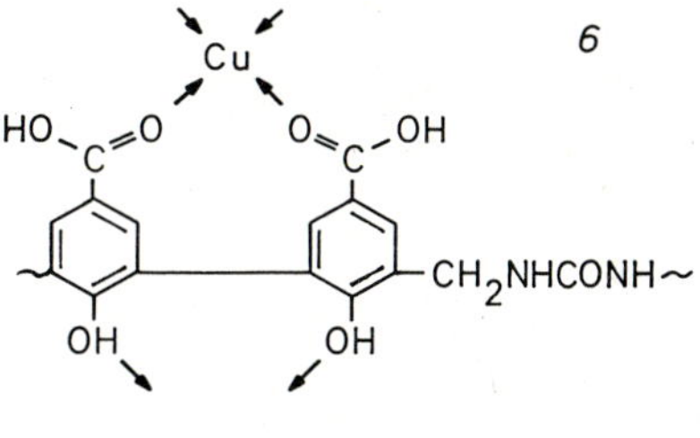

The assessment of the chelating ability of macroligands and applicability of such systems requires the data on the quantitative evaluation of the chelation processes such a stability constants and formation function. Contemporary methods are largely based on the composition of the products formed. The major methods of analyzing the quantitative parameters of MX_n chelation use predominantly the same techniques that are applied to the description of complexing reactions with participation of monofunctional macroligands [5, 7b, 13a].

Two extreme cases should be considered in the analysis of the quantitative parameters of the chelation processes: (1) the reaction occuring in the homogeneous

MX_n — macroligand systems (linear homo- and copolymers) and, (2) when the polymer ligand is suspended in the reaction medium.

First, let us discuss the chelation process in dilute solutions. Due to the fact that in this case the polymer chain is relatively long (except for chains of low molecular weight) the concentration of the functional end groups can be ignored. Considering a multiplicity of the reaction sites, two approaches to the study of the interactions in the MX_n — chelating macroligand (CML) systems have been adopted [27–30]. They differ in that in one approach the central particle is represented by a macroligand and in the other — by a metal ion. The first method consists in the examination of a successive addition of a transition metal to the functional groups of polymer. The method is based on the idea of an independence of every successive addition to chains containing different numbers of the already attached metal ions (similar to low-molecular weight analogues [31], the equilibrium constant being independent of the molecular weight of polymer. In the case of successive addition the role of a central particle is played by the macroligand; this can be written down as follows:

$$LM_{i-1} + M \overset{K_i}{\rightleftarrows} LM_i,$$

where LM_i is the chain containing i added M's.

Then

$$K_i = \frac{[LM_i]}{[LM_{i-1}][M]}.$$

The total stability constant $\bar{K}$ expressed through the current chain and metal concentration, [L] and [M], respectively, will be as follows:

$$\bar{K} = \frac{[LM_i]}{[L][M]^i} = \prod_{j=1}^{j=i} K_j.$$

The equations of material balance from the initial concentrations of the components, $[L]_0$ and $[M]_0$, have the form

$$[L]_0 = [L] + [L] \sum_{i=1}^{i=N} K_i [M]^i,$$

$$[M]_0 = [M] + [L] \sum_{i=1}^{i=N} K_i [M]^i,$$

where N is the number of the monomeric units of chain. Then the formation function ñ is determined as

$$ñ = \frac{\sum_{i=1}^{i=N} i K_i [M]^i}{1 + \sum_{i=1}^{i=N} K_i [M]^i}.$$

The successive reactions method is a modified version of the classical Bjerrum, and particularly Gregor, technique. It is based on the assumption of a uniform and independent distribution of the functional groups in volume. The concentrations of the functional groups and the complex are related to the total volume. However, this method is inapplicable in most cases due to some specific features typical of macroligand chelation, such as a high local concentration of the chelating groups in the coil, changes in the charges and conformation of the macromolecules in the course of the reaction, participation of sites belonging to the various chains and inaccessibility of part of the chelating groups for bonding. To account for these factors when calculating the stability constants, the concentration of the units capable of being bound with MX_n should be corrected. Thus it was proposed [32] that the concentration of the functional groups be related to the polymer volume; however, it is likely to vary during chelation to give rise to some experimental problems [33]. It should be noted that despite the obvious shortcomings of the successive reactions method, it is widely applicable at present in calculating the equilibrium constants for chelation reactions involving polymeric ligands.

The second method is based on Flory's principle of the independence of the reactivity of the bonding sites is independent of their position in the long chain or a low-molecular weight analogue with a correct selection of the model reaction components (a central particle in the form of a metal ion). For a single-site bonding the $\bar{K}$ value is determined as in the case of low-molecular weight compounds

$$\bar{K} = \frac{[M]_b}{([L]_0 - [L])[M]},$$

where $[M]_b$ is the total concentration of the complex, $[L]_0 - [L]$ is the concentration of the unreacted chain units.

Sometimes the chelation process is characterized by the substitution constant defined as the equilibrium constant ($\lg K_{eq}$) for the following reaction

$$2\,HL + M^{2+} \underset{}{\overset{K_{eq}}{\rightleftarrows}} ML_2 + 2\,H^+.$$

Table 1 lists as examples the chelation characteristics for the transition metal ions with poly(methacroyl acetone) (PMA), poly(vinylacetonyl ketone) (PVAK), acetyl-

Table 1. Stability constants of metal complexes with some β-diketones

Ligand	Cu		Ni		Co		Mn	
	$\lg K_{eq}$	$\lg K_2$	$\lg K_{eq}$	$\lg K_2$	$\lg K_{eq}$	$\lg K_2$	$\lg K_{eq}$	$\lg K_2$
PMA	−0.6	22.8	−11.4	17.4	−11.7	17.1	−13.7	15.1
PVAK	−4.8	7.0						
acacH	−2.7	18.4	−7.6	13.5	−8.5	12.6	−10.6	10.5
PDA	−4.8	8.0	−5.2	7.5	−5.4	7.3	−5.7	7.1

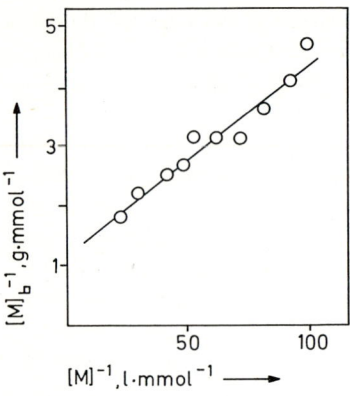

Fig. 1. Adsorption isotherm of the Hg^{2+} ion by macroligand with catechol groups

acetone (acacH) and poly (diethyl azelate) (PDA) [34]. Analysis of these data indicates that the CML (except for PMA) give rise to the complexes the stability of which is lower than that of the low-molecular weight analogues.

In studies of chelation involving insoluble polymers consideration should be given both to the diffusion (especially after the system has passed through the glass transition point T_g), and topological (practically a complete absence of the translation diffusion of the functional groups grafted to the polymer network) problems. Even the earlier studies [35] have shown that chelation with the use of CML is limited by diffusion. This leads from the fact that the surface chelating groups are the first to react whereas the penetration of MX_n into the block is restrained. Moreover, complexes appear frequently between the network nodes to result in an incomplete involvement of the functional groups into the process and dependence of the conversion Θ^1 on the reaction conditions. However, at small Θ's the concentration of the inactive and unreacted sites can be ignored. In this case the bonding constants can be estimated by a modified Langmuir equation [36]

$$1/[M]_b = 1/f_{max} + 1/(K[M]),$$

where [M] and $[M]_b$ are the equilibrium MX_n concentrations in solution and polymer-coordinated, respectively; f_{max} is the constant describing the limited bonding (adsorption) of the metal with the functional groups of polymer, i.e., $[M]_b/[L]_0$. As example the adsorption isotherm of the Hg^{2+} ions by macroligand with catechol groups is represented in Fig. 1. In this case the calculation [37] leads to the following results: $K = 42.3 \, l \, mmol^{-1}$ and $f_{max} = 0.86 \, mmol \, g^{-1}$.

2.3 Chelate Effect in Metallopolymeric Systems

By analogy with the chelate effect [38] its polychelate variety can be defined [39] as a gain of free energy ($\delta \Delta G_{pchel}$) in the addition of a CML to the metal ion

---

[1] Conversion is the ratio of the actual number of metal ions bound onto a chain to the maximum possible one.

($\Delta G_{p\,chel}$) as compared with the case when the monodentate ligands with the same donor atoms are added to such an ion (ΔG_m)[2]

$$\delta \Delta G_{p\,chel} = \Delta G_{p\,chel} - \Delta G_m.$$

However, in contrast to LMWM, the quantitative parameters of chelation involving macromolecular ligands have not been studied well enough. This is associated with a need to consider a number of additional factors controlled by the chain type of the ligand. It is apparent that for PCMU the following three levels of their spatial arrangement should be considered:

1) a local level that represents a chemical structure of a local chelate node (type of the complexing metal, donor atoms, cycle size, spatial arrangement of the cycle, etc.);

2) a molecular level due to the chemical structure of the polymer chain (length of the chain, elemental composition of the repeating units, shape and conformation of the chain, etc.);

3) a supramolecular level representing a type of the intermolecular interaction of macromolecules and the degree of their mutual ordering.

Considering the three levels a change of the free energy when chelation involves macroligands (assuming its additivity) is as follows:

$$\Delta G_{p\,chel} = \Delta G_{loc} + \Delta G_{mol} + \Delta G_{supr},$$

where ΔG_{loc}, ΔG_{mol} and ΔG_{supr} are the changes of free energy for the local, molecular and supramolecular levels, respectively. The equation for the polychelate effect has the following form:

$$\delta \Delta G_{p\,chel} = \Delta G_{loc} + \Delta G_{mol} + \Delta G_{supr} - \Delta G_m.$$

Due to the fact that the difference $\Delta G_{loc} - \Delta G_m$ represents a local chelate effect ($\delta \Delta G_{l\,chel}$),

$$\delta \Delta G_{p\,chel} = \delta \Delta G_{l\,chel} + \Delta G_{mol} + \Delta G_{supr}.$$

Thus, as compared with the chelate effect, its polychelate form will include two additional terms for the molecular and supramolecular levels of the spatial PCMU arrangement.

Under some conditions, the free energy changes for a given level can be ignored, so that it is possible to carefully analyze the main contributions to ΔG for such a level. First, we shall consider dilute polymer solutions in which their association can be neglected. Then $\Delta G_{supr} \to 0$ and, hence

$$\delta \Delta G_{p\,chel} = \delta \Delta G_{l\,chel} + \Delta G_{mol}$$

---

[2] This definition is valid only in the case of the correctly selected components of the model reaction.

In the case of infinitely long chains ($l \to \infty$) and small conversions ($\Theta \to 0$) the chelation process has no appreciable effect on the macromolecule shape and conformation, therefore $\Delta G_{mol} \to 0$ and

$$\delta \Delta G_{p\,chel} = \delta \Delta G_{l\,chel}$$

Thus for dilute solutions at $l \to \infty$ and $\Theta \to 0$ the chelating ligand behaves like a low-molecular weight analogue so that it is possible to ignore the molecular and supramolecular levels and consider only the local chelate nodes. As the nature of the complexing metal and donor groups in the two cases is the same, the chelation enthalpy is little affected when passing from the LMWM to the PCMU. For example, one can observe similar values of ΔH in the chelation of copper(II) ions with poly(amidoamines) of the types **7** and **9** and their low-molecular weight analogues of the types **8** and **10** (Table 2) [40]. This suggest a similar spatial structure of the chelate nodes in these compounds. Therefore for both the LMWM and PCMU the chelate effect will be controlled by the entropy component. This is indicated by the same type dependences of the ΔG and ΔS parameters in the chelation processes on conversion (Fig. 2) and the independence of the ΔH of Θ (Table 2).

It should be noted that the entropy component of the polychelate effect is largely controlled by the same contributions as for LMWM, i.e., translation, rotation, symmetry, isomerism, oscillation, internal rotation and solvation contributions [41]. Besides both the chelate and polychelate effects depend greatly on the size of the chelate cycle. In this case the PCMU obey the known Chugaev rule by

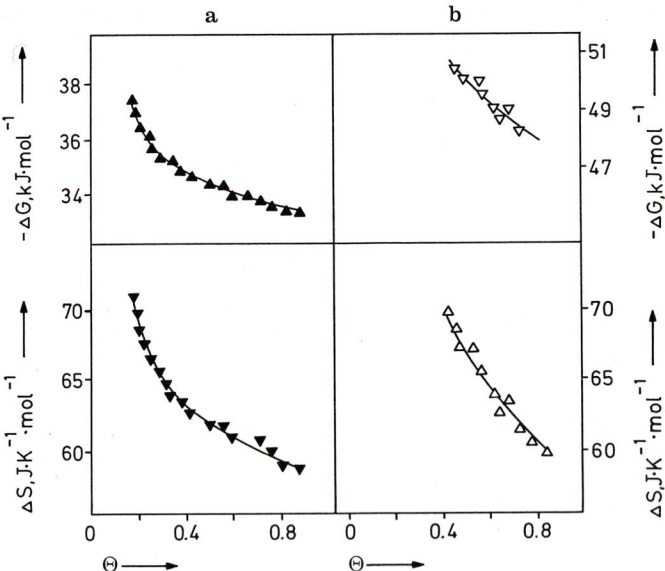

Fig. 2. Dependence of the thermodynamic functions at 25 °C in 0.1 mol l^{-1} NaCl on the conversion for (*a*) the polymer **9** – Cu(II) system (2:1 molar ratio), and (*b*) the polymer **7** – Cu(II) system (1:1 molar ratio)

Table 2. Thermodynamic functions of copper(II) complex formation

Ligand		$-\Delta G$ (kJ·mol^{-1})	$-\Delta H$ (kJ·mol^{-1})	ΔS (J·K^{-1}·mol^{-1})
~C-N⌒N-CCH$_2$CH$_2$NCH$_2$CH$_2$~ ‖ ‖ | O O CH$_2$COOH	(7)	51.5–47.3^a	30	72–58^a
O⌒N-CCH$_2$CH$_2$NCH$_2$CH$_2$C-N⌒O ‖ | ‖ O CH$_2$COOH O	(8)	57.0	30	90
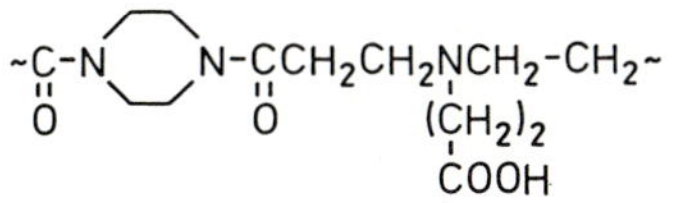 ~C-N⌒N-CCH$_2$CH$_2$NCH$_2$-CH$_2$~ ‖ ‖ (CH$_2$)$_2$ O O COOH	(9)	37.2–33.1^b	16	71–58^b
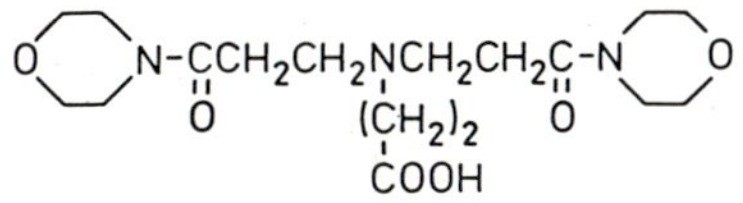 O⌒N-CCH$_2$CH$_2$NCH$_2$CH$_2$C-N⌒O ‖ (CH$_2$)$_2$ ‖ O COOH O	(10)	34.5	17	58

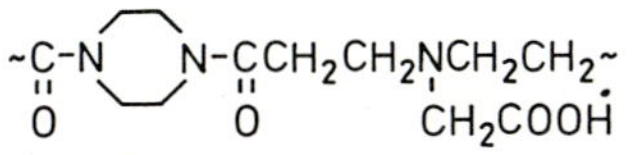

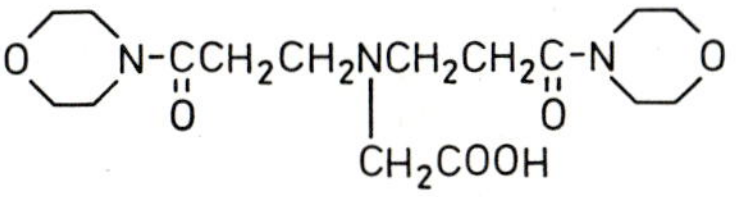

^{a} $\Theta = 0.45$–0.85;
^{b} $\Theta = 0.18$–0.88

which five- and six-membered cycles are stable. The effect of cycle size on the mechanism of bonding metal ions with polymers can be exemplified by TiCl$_4$ complexing with the copolymers of styrene (St) and diallyl dicarboxylates [42]. At $t = 1$ or 2 one can observe the appearance of mixed complexes with cis-(11) or trans-(12) position of the carbonyl groups:

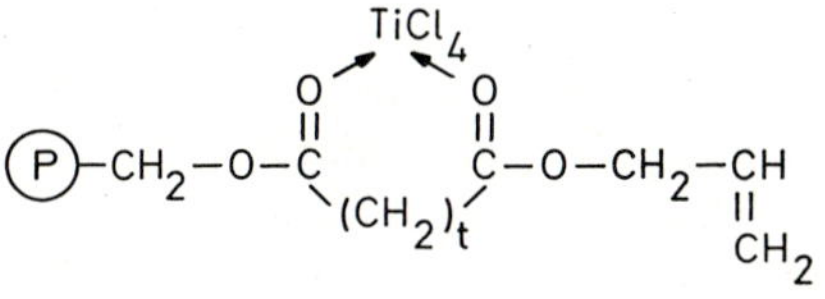

$$\text{(P)}-CH_2-O-\underset{\underset{(CH_2)_t}{}}{\overset{O}{\overset{\|}{C}}}\overset{TiCl_4}{\overset{\searrow\swarrow}{}}\underset{}{\overset{O}{\overset{\|}{C}}}-O-CH_2-\underset{\underset{CH_2}{\|}}{CH} \qquad 11$$

$$\text{(P)}-CH_2-O-\overset{\overset{\uparrow}{\overset{O}{\overset{\|}{TiCl_4}}}}{C}-(CH_2)_t-\underset{\underset{TiCl_4}{\downarrow}}{\overset{\overset{O}{\|}}{C}}-O-CH_2-\underset{\underset{CH_2}{\|}}{CH} \qquad 12$$

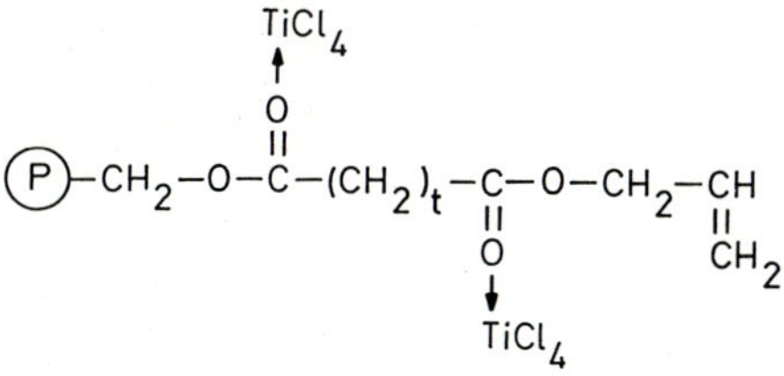

where the symbol Ⓟ denotes the polymer chain. Two dissimilar carbonyl groups exists in both the parent ligand and the PCMU, although the difference between

$\Delta\vartheta_{C=O}$ values for the *cis*- and *trans*-forms in this case increases from 10–15 to 40–45 cm^{-1}. This suggests the existence of a more rigid fixation of the isomers in the course of complexing. Separation of ester groups (t > 3) renders chelation impracticable. However, the polymer chain is responsible for the spatial structure of the chelating fragment, and the metal ion "selects" the most preferable conformations on the macromolecular chain. Therefore metallocycles different from those in low-molecular weight analogs can be formed. For example, in the interaction of copper(II) ions with vinyl polymers containing aminoalkylamide units in the side chain, one can observe the formation of seven-membered cycles. An amine nitrogen atom and an amide C=O group participate in metal coordination [43]:

$$\begin{array}{c} \text{P} \\ | \\ \text{C=O} \\ \text{HN} \diagup \quad \searrow \\ | \quad\quad\quad \text{Cu}^{2+} \\ \text{H}_2\text{C} \diagdown \quad \nearrow \\ \text{CH}_2-\text{NH}_2 \end{array}$$

At the same time, five-membered cycles are formed in the low-molecular copper(II) – glycilglycine system [44]. Thus a knowledge of the local chelating fragment conformation allows us to predict PCMU with a specified size of the metallocycles.

Along with the contributions to the chelation entropy and enthalpy common to the low- and high-molecular weight metallochelates, the PCMU are distinguished by some specific contributions. One has to consider a contribution associated with the "adjustment" of the local chelating fragment since the interaction of the metal ions with the CML is accompanied, along with a direct chemical act (formation of a metallocycle), by a change in the "local rigidity"[3] in the reaction site due to a variable interaction of the valence-unbound chain atoms. The energy of such an "adjustment" for soluble macroligands is far lower than that of chelation therefore the structure of the resulted coordination site is normally controlled by the nature of a complexing metal. For instance, in the interaction of iron(III) salts with polystyrene containing acac groups the "adjustment" of the polymer chain gives rise to octahedral structures without particular strains [13c]. In addition, a significant effect of the "local rigidity" changes on the stereochemistry of the resulted chelate nodes is not excluded, either. It is known [46] that bond lengths are less variable than bond angles and therefore, in PCMU, the latter are expected to suffer stronger deformation than the former. Moreover, the PCMU show the presence of a contribution caused by the "neighbour groups" effect due to the existence of both reacted and unreacted units which influence the free energy of the chelate fragment formation and, hence, the chelate node structure. This is attributed mainly to the steric factors which result in the repulsion of the chelate nodes, their deformation and chelation constraints with increased conversion. Thus, for example, the octahedral complexes of iron(III) ions with hydroxamic

---

[3] The "local rigidity", as defined elsewhere [45], is a bending resistance of a uniform restilinear filament.

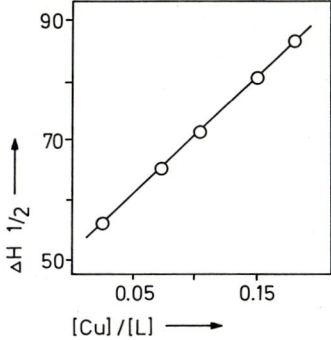

Fig. 3. Relationship between width at half-height ($\Delta H_{1/2}$) of the low-field component of the ESR spectrum of copper complex with poly(hydroxyphenylbenzoxazolterephthalamide) and the ratio [Cu]/[L] at $-196\,°C$

acid copolymers are formed readily [47] only in the case when the chelating fragments are spaced by eleven atoms. At k = 9 (k is the number of atoms between hydroxamic groups) there appear small strains and at k = 3 the formation of such complexes is altogether impracticable. The chelating fragments located side-by-side in the polymer chain inhibit the exchange interactions between the metal ions. Thus a greater number of copper(II) ions bound with poly(hydroxyphenyl benzoxazole therephthalamide) give rise to broadened ESR spectral lines due to the dipole-dipole interactions between the paramagnetic centers (Fig. 3) as a result of shortened distances between them [48]. However, even at a maximum bonding of the copper(II) ions with a polymer the spectrum does not become singlet-type which suggests the absence of a strong exchange interaction between metal ions. The cooperative effects to show up [49] at distances of 60–90 nm, hence, the minimum distances between the adjacent copper(II) ions in such PCMU exceed this value, as distinct, for example, from copper — linear poly(ethylene imine) complexes with no steric hindrances for the exchange interactions between the metal ions [50]. At the same time, the separation of the chelating fragments brings about unfavourable entropy factors. In particular, when passing from the LMWM to the PCMU based on picolinic acid and its macromolecular analogue (a styrene — vinyl picolinic acid copolymer) one can observe a decrease in the stability constant (Table 3) [51]. If the chelating fragment is situated in the side chain, the effect of the main chain will depend on the length of the spacer bridge and with increasing length of this chain the difference between the chelate and polychelate effects disappears. Thus, for example, for the PCMU of the type

$$\text{(P)}-(CH_2)_m-NH-\underset{\underset{Fe/3}{O\quad O}}{\overset{}{\bigcap}}-CH_3$$

above m = 2 the complexes are equally stable [52]. If, on the other hand, the polymer creates difficulties in the formation of chelate complexes (as in the case of polystyrene with the Dipy groups fixed with respect to the 6-position of the

Table 3. lg K values of metal complexes with picolinic acid and its polymeric analogs (dioxane-water)

Ligand	Cu(II)	Ni(II)	Co(II)	Zn(II)	Mn(II)
Picolinic acid	12.80	12.70	12.50	12.72	9.50
Polymeric analog	9.88	9.56	8.28	8.44	8.44

benzene ring [53]) the metal will be bound in a monodentate manner:

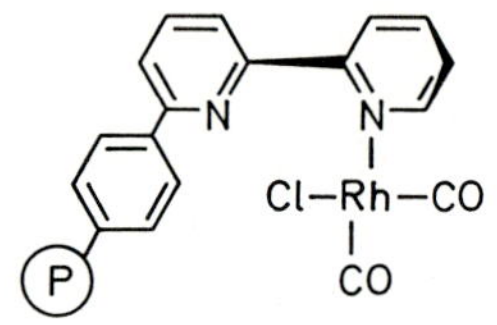

The macroligand permits us to trace the stagewise mechanism of PCMU formation, to isolate the intermediate and stabilize (in the case of unstable low-molecular weight analogs) complexes.

Thus, at the local level the polychelate effect is close to the chelate one provided the model reaction components are selected correctly. However, in doing so one has to account for the polymeric nature of chelating ligand and the associated contributions to the chelation entropy. In the reaction

$$[CuX]^+ + Y^- \rightleftarrows [CuY]^+ + X^-,$$

where Y^- is the poly(amidoamine), X^- is its low-molecular weight analog, the transfer contribution will be equal to zero (the number of particles remains the same). A free energy change can be obtained from Eq. (1) and determined from Eq. (2):

$$\Delta(\lg K) = (\lg K_{[CuY]^+}) - (\lg K_{[CuX]^+}), \tag{1}$$

$$\Delta G = -2.305RT \, \Delta(\lg K) = -A_{el} - TS_{st}, \tag{2}$$

where A_{el} is the loss of the electrostatic energy in the transfer of the cation charge from the negative charge on the polymer into infinity, S_{st} is the statistical factor. The following Eqs (3) and (4) are used for calculation of the change of the enthalpy and entropy of the ligand exchange process:

$$\Delta H = -A_{el} \left[1 + \frac{T}{\varepsilon_{ef}} \frac{\delta \varepsilon_{ef}}{\delta T} \right], \tag{3}$$

$$\Delta S = -A_{el} \left[\frac{1}{\varepsilon_{ef}} \frac{\delta \varepsilon_{ef}}{\delta T} \right] + S_{st}, \tag{4}$$

Table 4. Thermodynamic functions for ligand exchange reaction

$\Delta(\lg K)$	$-\Delta G$ (kJ/mol)	TS_{st} (kJ/mol)	S_{st} (J/K · mol)
0.65	3.72	3.72	12.5
0.45	2.55	2.55	8.6
0.31	1.76	1.76	5.9
0.22	1.26	1.26	4.2
0.06	0.33	0.33	1.1
0.01	0.04	0.04	0.1
−0.02	−0.13	−0.13	−0.4
−0.10	−0.59	−0.59	−2.0
−0.13	−0.75	−0.75	−2.5
−0.20	−1.13	−1.13	−3.8
−0.40	−2.30	−2.30	−7.7
−0.49	−2.80	−2.80	−9.4
−0.73	−4.18	−4.18	−14.0

where ε_{ef} is the effective dielectric constant. As could be expected, the ΔH value for this process is equal to zero [40] because the same type of the ligand environment of the copper(II) ions in the LMWM and PCMU. The entropy component decreases with increasing conversion from positive to negative values (Table 4). It reaches zero, i. e., the state when the chelate and polychelate effects are the same. A half of the negative charges on the polymer is linked with the copper(II) ions or, in other words, when the positive and negative charges on the polymer are equal to one another. Further addition of such ions leads to the situation when positive charges on the polymer predominate so that a back reaction occurs.

At a higher conversion and shorter polymer chain length one has to account for a molecular PCMU level since these parameters produce a significant effect on the shape and conformation of macromolecules in solution. Quite frequently, the chelation allows the stabilization of the conformational and tautomeric forms not typical of the parent polymer. In particular, noticable changes in the shape of poly(2-vinylpyridine) have been detected [54] in its interaction with $Co(acac)_2$. Conformational transitions were observed [55] in linear poly(amidoamines) in the course of protonation and complex-formation in that participation of the main chain nitrogens involves conformational transitions affecting the properties of the macromolecule as the sites where the various monomeric units (six-membered chelate diacylpiperazine cycles) are joined together are very rigid. The addition of $[Fe(2,2',2'',2'''-terpyridyl)]^{3+}$ complex ions to Na-poly(L-glutamate) solutions increases the amount of the α-helical form at thos pH values when usually a coil shape predominates [56]. In the interaction of copper(II) ions with poly-N^5-(3-hydroxypropyl)-L-glutamine (PHPG) the α-helical conformation predominates up to $\Theta = 6.25 \times 10^{-2}$ and increasing this value results in chain conformation changes (Fig. 4) [57].

When passing from a dilute to a concentrated solution or solid phase, one can observe a stronger intermolecular interaction of both the initial polymers and the

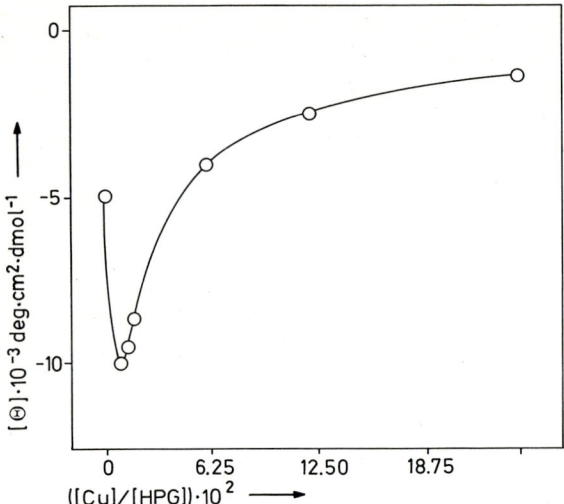

Fig. 4. Ellipticity ($[\Theta]_{max}$) values of Cu(II) — PHPG complex versus [Cu]/[HPG] ratio at pH = 10 and [HPG] = 5×10^{-4} mol l^{-1}

derived metallochelates. Therefore it is necessary to consider the third level of the spatial PCMU arrangement. The parameters of such an interaction are affected by the donor ability of the functional groups which controls the stability of the complex and the strength of the interaction, the distance of functional groups from the main chain as well as the presence and size of the substituents. These factors are responsible for the situation when polymer complexes frequently form more perfect morphological structures than the parent macroligands. Thus, detailed studies were devoted to the structural changes of styrene — divinylbenzene copolymer (St-co-DVB) when alkylated with 5-chloromethyl-8-hydroxyquinoline in nitrobenzene in the presence of $AlCl_3$ [17c]. The complex undergoes micronetwork rearrangements and twisting. The structural changes obey the steric requirements of the complexing metal (minimum losses of free energy) and the susceptibility of the molecular chains to twisting is dependent on chain rigidity and type of solvent. These factors are capable of weakening or, on the contrary, strengthening the chain twisting or extension.

As the structure of complexes is greatly affected by the conditions of their preparation, one can expect the formation of thermodynamically nonequilibrium products for which one has to consider the relaxation time and, hence, the dynamics of changing the desired properties. In rigid-chain polymers the appearing chain-to-chain cross-links appreciably change their supramolecular organization to result in stiffened chains, which is expected to be displayed by changes in the relaxational transitions within the chains and their segments. It is apparent that the greatest effect is experienced by α-relaxation associated with the micro-Brownian motion of the segments near T_g. At low degrees of polyligand filling the total limitation of chain mobility is insignificant whereas at higher degrees, a larger number of cross-links between the adjacent chains and thereby higher T_g values. At the same time β-relaxation is practically independent of metal content. During cyclization at higher metal content increases T_g due to the stiffening of segments and lowering of the mobility of the chain.

It should be noted that conversion changes may bring about a reversible transition between the intra- and intermolecular metallochelates. This can be illustrated by the following scheme of iron(III) ions — polyhydroxamic acids complexing [58]:

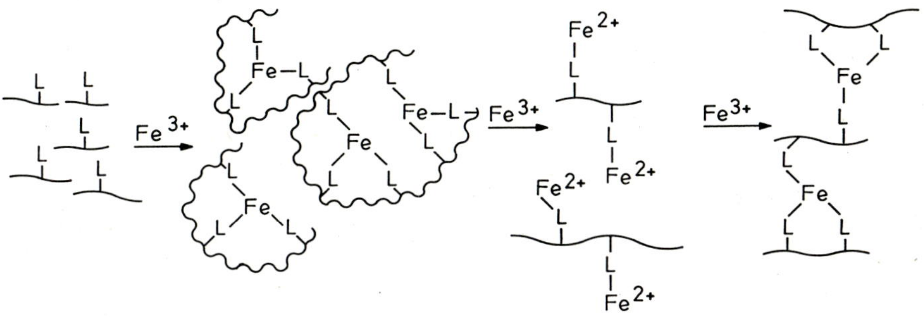

The first stage, at small $[Fe^{3+}]/[L]$ ratios, involves the formation of 3:1 complexes with intramolecular bonds. A further increase of this ratio yields 1:1 complexes but also with the same bond type. At high conversions there exist transitions from the intra- to intermolecular 3:1 metallochelates which are insoluble and are precipitated from solution.

3 Preparation of Polymers Containing Metallochelate Units

Modern synthetic PCMU chemistry uses broadly the general methods and principles of complex compound synthesis [59]. The currently available methods for the preparation of such polymers are shown in Scheme 1. As in the case of LMWM most PCMU are obtained by the direct interaction of metal ions with CML. At the same time, PCMU chemistry also relies on some other methods, particularly "assemblage" of metallochelate nodes.

3.1 Syntheses Using Chelating Macroligands

Depending on the nature of the base polymer, chelation reactions can occur either in a homogeneous (soluble linear homo- and copolymers) or heterogeneous (cross-linked, grafted polymers and gels) phases. Consideration of the polymeric nature of the chelating ligand requires a more deliberate, as compared with the LMWM, selection of solvents, pH control and specific synthesis conditions.

Selection of the metal compound is dictated by a number of requirements. First, the availability of the metal compound. As indicated above, PCMU have been obtained with the use of nearly all the metals of the Periodic Table. For many metals a great number of salts are known whereas for some ions only a few compounds are stable e.g. VCl_4, largely in the inert atmosphere. Salts with the

Scheme 1

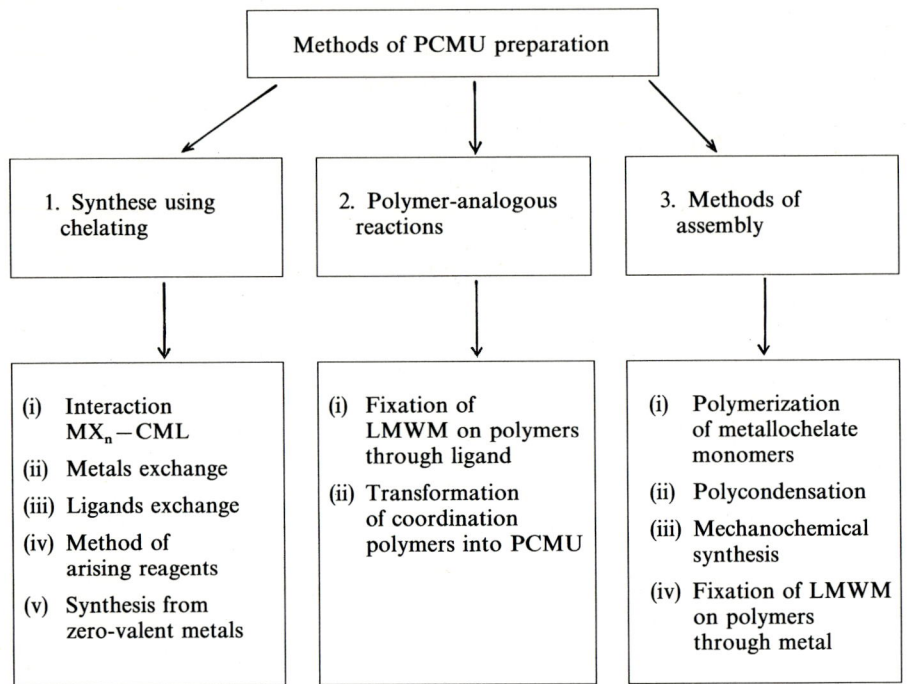

highest stability and availability are preferred. The second requirement is critical in the case of soluble macroligands as it is frequently desirable that the ligand and metal salt be soluble in the same solvent. For insoluble polymers, this condition is associated with the ability of the ligand to swell in the same solvent that is used to dissolve the metal salt. Third, a metal salt is selected with regard to the type of the metallochelate produced. Thus molecular metallochelates are obtained with the use of the same metal compounds that is employed for the preparation of a given metallocomplex. In this respect metal halides are the most frequently used species [10a, 11b–d, 12], because of their comparative availability and stability, the same is also true for sulfates and nitrates [11a, e]. For intracomplex compounds the correct selection of the salt is especially important in that it defines not only the pH of the medium, which is essential for the synthesis of the complexes, but also the probability of various complications during the synthesis. Most studies have been made with metal acetates since they are the most favourable for the synthesis of complexes involving a variety of organic compounds whose chelate H-cycles contain a number of heteroatoms [59a, b, 60, 61]. For example, if the chelation process with a polymer containing a β-diketone involves copper sulfate, nitrate and acetate under the same conditions (26 °C, 20 h), the degree of chelation is 18.9, 20.5 and 94.4%, respectively [62].

The nature of the solvent plays a greater role in the synthesis of PCMU than LMWM, which is attributed to the polymeric nature of the chelating ligand. The solubility of the polymer is poorer than that of its low-molecular weight analog, this difference becoming greater with increasing molecular weight of the polymer. Therefore, the selection of a solvent equally suitable for both a macroligand and metal salt is quite a problem. For soluble polymers the starting species is represented by a CML solution in an organic solvent (e.g. dioxane, acetone [13a, b], DMF [11e, 13a, b, d, 14a–c, 63–65], THF [3a, 11e, 12, 13d, 24]) or mixed media (e.g. water-DMF [66]). The synthesis includes mixing of the reagents with the subsequent extraction of the macrocomplex with the use of a selective solvent. In many situations the macrocomplex precipitate can be obtained during the synthesis [3a, 14a–c, 17a, 24, 66, 67]. If one fails to find solvents that are specific with respect to the two reagents, PCMU can be produced at the interface of both immiscible liquids. Thus, for example, a complex of copper(II) with a polymer containing hydroxamic acid groups was obtained by interaction of a ligand solution in cyclohexane and an aqueous solution of copper acetate [68]. For comparison the same complexes were obtained in a homogeneous medium with the use of DMF as the solvent. The results of such studies indicate that the interphase method is even more preferred as compared with the homogeneous one (Table 5). In the case of insoluble polyligands, the most typical method — a suspension reaction — has its advantages and disadvantages. Here the most important thing is the selection of solvent for macroligand swelling. If the polymer does not swell in a given solvent, chelation either does not take place [69] or a small percentage of metal is bound [70].

pH of the medium plays a critical role in synthetic PCMU chemistry as it defines the type of metal complex. Depending on pH, copper(II) ions are shown to give a number of complexes with poly(amidoamines) such as CuL^{2+}, $Cu(OH)L^+$, $Cu(OH)_2L$ [71]. In the case of chelation, especially in water, the pH of the reaction mixture should not exceed the value at which metal hydroxides are formed. In the preparation of PCMU such a value should be controlled during the synthesis. Of particular importance is the pH value in the preparation of intracomplex compounds most of which are obtained in slightly acid, neutral or slightly alkaline media. Moreover, due to the fact that most PCMU are low soluble compounds, the problem of pH control becomes still more critical. Figure 5 depicts X-ray diffractograms for copper(II) — poly(*o*-isophthaloylphthalamidoxime) complexes

Table 5. Copper(II) chelation with polymers containing hydroxamic acid groups

System	Temperature, °C	Time, h	Copper content, %	Chelation degree, %
Homogeneous	20	50	3.62	36
Homogeneous	50	50	4.96	49
Interphase	20	50	5.94	59
Interphase	40	50	7.05	70
Interphase	60	50	10.08	100

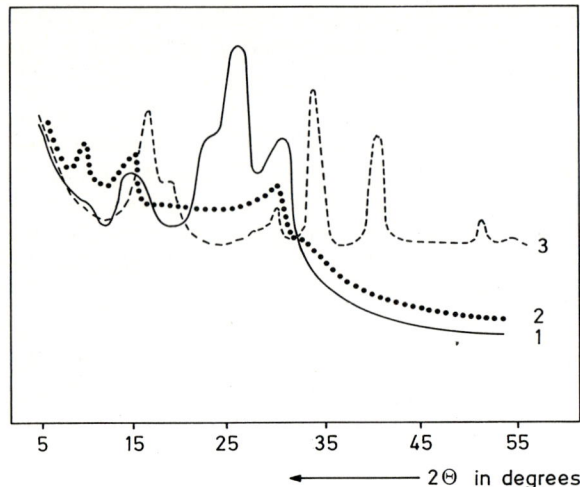

Fig. 5. X-ray diffraction patterns of poly(o-isophthaloyl-phthalamide oxime) (1) and its complex with copper(II) ions (2). For comparison, X-ray diffraction pattern of the same complex containing copper hydroxide as impurity is included (3)

which can be used as effective indicators for the presence of hydroxides in a complex as even small amounts of impurities have a great effect on the diffractograms [72].

Increasing temperature, favors a greater polymer yield. This is shown by the data in Table 5. At the same time, the temperature should be strictly controlled, especially considering the fact that many organic polymers are extremely unstable at elevated temperatures. Moreover, it should be taken into account that increasing temperature may lead to:
i) complex decomposition;
ii) complexes with unexpected qualitative and quantitative compositions;
iii) facile side processes, especially hydroxide precipitation.

The ratio of reagents affects dramatically the composition of the complexes resulted. In this respect PCMU differ greatly from LMWM which in most cases are distinguished by the formation of complexes with the same composition irrespective of the ratio of the starting reagents. In PCMU chemistry, however, such a relationship has a marked effect on a variety of characteristics of corresponding complexes, such as composition, solubility, etc.

The time factor of various methods suitable for PCMU synthesis is also a characteristic which ranges between 1 [10a, 73] and 24 hr [11b, 13c, d, 14f, 24, 74, 75] (sometimes even longer [11c, d, 17c]). Also it should be noted that increasing reaction time may bring about various reactions, particularly reduction of the transition metal (Fig. 6) [76].

In addition to a direct interaction of MX_n with chelating polyligand PCMU can be obtained by exchange of metals [22d, 77]. This method consists in mixing MX_n with polymeric chelate compounds, mainly, incorporating alkaline and alkaline-earth metals. In this case the first stage consists of the preparation of the corresponding metal derivative with the subsequent synthesis of PCMU. This can be exemplified by the synthesis of a nickel(II) complex with the product obtained

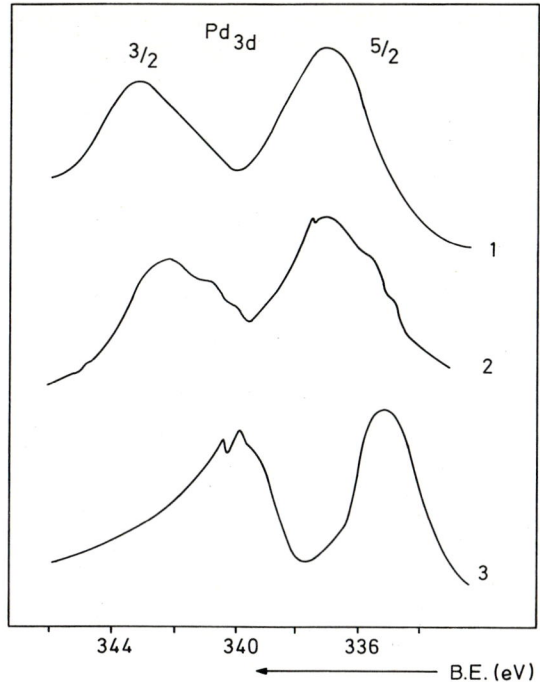

Fig. 6. Palladium 3d photoelectron lines of palladium acetate (*1*), and complexes prepared by the interaction of palladium acetate and polymer with Dipy groups for 5 (*2*) and 24 h (*3*)

by interaction of polyethylene-graft-poly(methylvinyl ketone) (PE-g-PMVK) with *o*-aminophenol [77c]:

where ▨ denotes the polyethylene surface. Another alternative is the formation of a metal derivative by macroligand synthesis [22d]:

Due to the fact that the preparation of most intracomplex compounds requires neutralization of the acid formed in the reaction, which is undesirable because of the formation of metal hydroxide, the metal-exchange method appears to be an efficient tool for the production of just this type of PCMU.

Another interesting method for preparing PCMU from CML is that of ligand exchange based on the displacement of the low-molecular weight chelating ligand from the coordination node of the complex by the chelating fragment of the polymer. For example, the ligand exchange mechanism was used to obtain a nickel(II) complex with the product resulting from the interaction of PE-g-PMVK and salicylic acid hydrazide [14e]:

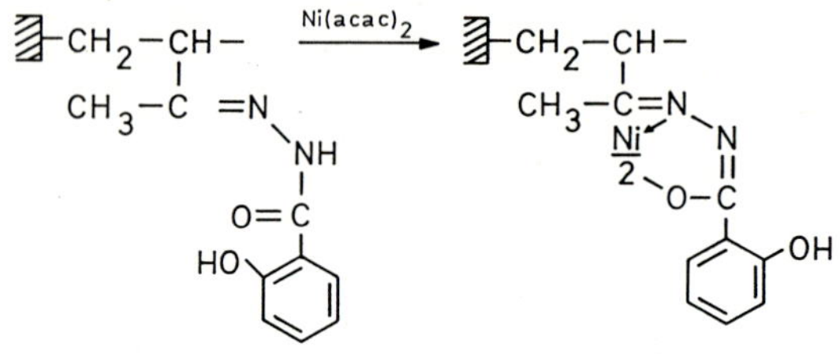

Unfortunately, a method of synthesis of complexes from zero-valent metals, used extensively in LMWM chemistry [78], has not been widely applied to PCMU chemistry although a few isolated attempts have been successful [79]. This permits the combination of stages at which MX_n is synthesized and bound to the polymeric chelating ligand. The main problem lies in the separation of the PCMU obtained from the unreacted metal.

3.2 Polymer-Analogous Reactions

The method of polymer-analogous reactions is mostly based on the fixation of LMWM on polymers through a ligand. This is effected by interacting the reactive peripheral groups of the LMWM with the functional groups of the polymer carrier. Carboxyl, sulfo groups or their sodium derivatives are the most commonly used as peripheral groups. This method of PCMU synthesis is one of the most widely used techniques used to immobilize transition metal porphyrins and phthalocyanines. A survey of the LMWM and macromolecular ligands used for this purpose has been published elsewhere [7b, e]. Immobilization of metal acetylacetonates on chloromethylated styrene − divinylbenzene copolymer (CMCS) proceeds quite readily too. For example, Ni(II) and Fe(III) acetylacetonates were fixed in the presence of KI [13e]:

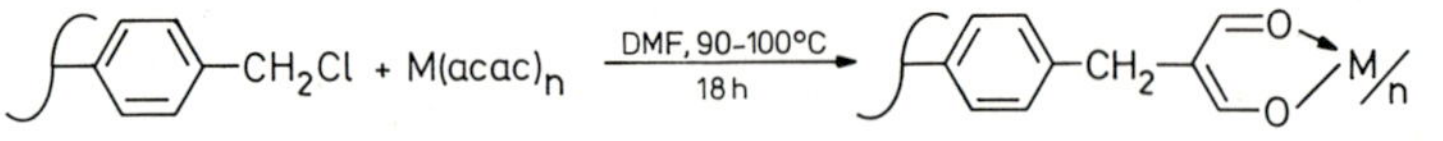

The interaction between tris(2,2′-dipyridyl)ruthenium(II) with poly(p-styrene sulfonate) [80] due to the outer-sphere cooperative processes yields the following structures:

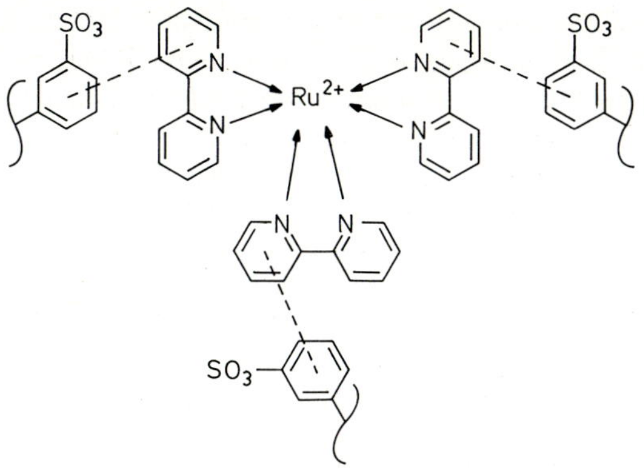

The method of conversion of coordination polymers into PCMU consists of the preparation of a coordination polymer and its subsequent conversion into PCMU. It was used to obtain metal complexes with Schiff polybases [81]:

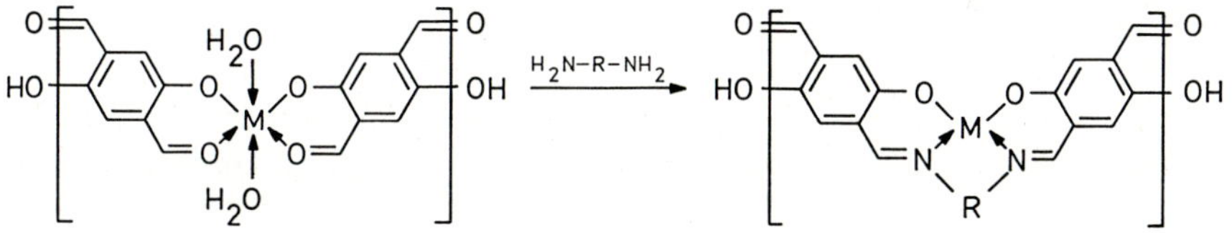

This method is disadvantageous in that the reaction should be carried out in a suspension (considering the insolubility of coordination polymers in organic solvents) and the probability of various side reactions occuring.

3.3 Methods of Assembly

Polymerization and copolymerization of metal chelate monomers (MCM) is a unique method for the synthesis of metallopolymers practically all of whose chelating fragments are linked to the metal [82]. Chelation principles are rather popular for stabilizing metallomonomers, particularly organometallic compounds. Thus allyltitanium trichloride is extremely unstable, however its complex with 2,2'-dipyridyl decomposes only at about 180 °C [83]. Unsaturated organocobalt compounds are largely stable in that case only when they include bulky chelating ligands, particularly dimethylglyoxime (DH_2) [84].

Initial advances have been made in MCM synthesis and further studies are in progress [82]. It was not until recently that polymerization kinetics of such systems and properties of the resulting PCMU have received attention. Thus, radical polymerization of the dimethyl ester of hemin was carried out in 1977 [85]. The possibility of polymerizing 4-vinylpyridine (4-VP) complex-bound to Ru(II) (cis-$[Ru(Dipy)_2(4-VP)X]^{n+}$ complexes, where n = 1, X = Cl; n = 2, X = CO or 4-VP) has been shown [86a] but the properties of the polymers obtained have not been described. It should be noted that homopolymerization of MCM does not proceed readily and in most cases it is precluded altogether. Besides the polymerization rate of the MCM is lower than that of its "metal-free" analog [87].

More detailed studies were devoted to the mechanism of electrochemical polymerization applied to Fe(II), Ru(II) and Os(II) complexes containing 2,2'-dipyridyl and monofunctional monomer (L) such as 4-VP, bis(4-pyridyl)ethylene, *trans*-4-stilbazole or *N*-(4-pyridyl)acrylamide [86]. The first stage of electrochemical polymerization is shown to be the formation of a radical-anion, e. g. by the following scheme

$$[Ru(II)(Dipy)_2(L)_2]^{2+} + \bar{e} \rightleftarrows [Ru(II)(Dipy)(Dipy^-)(L)_2]^+ + \bar{e}$$

$$\rightleftarrows [Ru(II)(Dipy^-)_2(L)_2]^0$$

The polymerization is induced by a one-electron reduction based on the intramolecular redox equilibrium including a thermodynamically less favourable isomer in which the electron is located on L:

$$[Ru(II)(Dipy)(Dipy^-)(L)_2]^+ \rightleftarrows [Ru(II)(Dipy)(Dipy^0)(L)(L^-)]^+$$

The copolymerization of MCM with tradititional monomers is a more popular process than their homopolymerization. Copolymers containing complex-bound compounds of rare-earth metals such as Eu(III) and Tb(III), were obtained by copolymerization of MCM of the type M(4-vinyl-4'-methyl-2,2'-dipyridyl)$_3$ with methyl methacrylate (MMA) in block or methanol [88]. Radical copolymerization was also effected in the case of complexes based on 4-VP and ruthenium compounds such as *cis*-[Ru(Dipy)$_2$(4-VP)Cl]ClO$_4$, *cis*-[Ru(Dipy)$_2$(4-VP)$_2$](ClO$_4$)$_2$ and *cis*-[Ru(Dipy)$_2$(4-VP)(CO)](PF$_6$)$_2$. Used as the comonomers were St, MMA and 4-VP. The reaction occured in toluene or acetonitrile. The copolymer of Ru(II) complexes containing a molecule of 4-VP turned out to be soluble in dichloromethane and methanol whereas the copolymerization product [Ru(Dipy)$_2$(4-VP)$_2$](ClO$_4$)$_2$ (comonomer is 4-VP, the initial ratio [4-VP]/[MCM] = 20) was insoluble in organic solvents due to the fact that the polymeric chains were cross-linked through the Ru atoms [86a]:

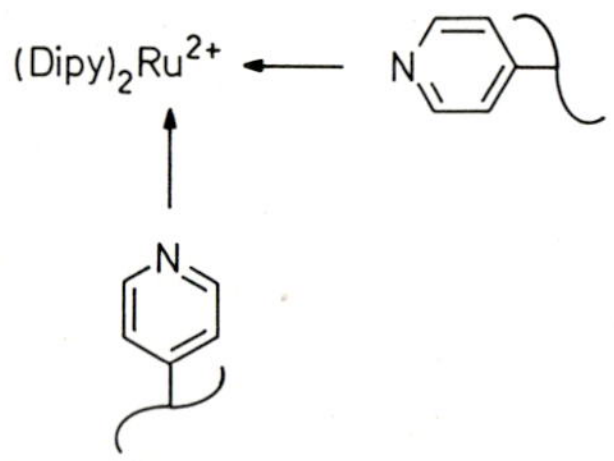

However, as the ratio increased to 200, the effect of MX$_n$ became smaller so that a soluble product was obtained.

It has been shown that *cis*-[Ru(Dipy)$_2$(4-VP)X]$^{n+}$ (n = 1, X = Cl; n = 2, X = CO or 4-VP) can polymerize with St or MMA [86a, e], although neither the mechanism of the process nor the properties of the resulted products have been

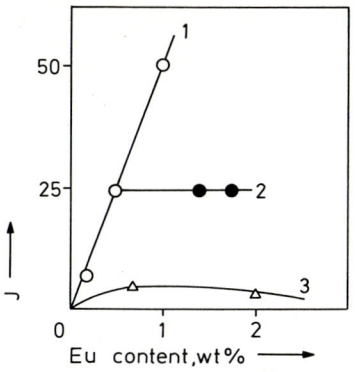

Fig. 7. Relationship between fluorescence intensity (I) and Eu content in complexes 13 (*1*), 14 (*2*), and 15 (*3*)

described. No mechanism for the copolymerization of MCM — Eu[2-(4'-vinylbenzoyl)benzoate]-di(2-benzoyl benzoate) with St to give product 13:

Interestedly that this metallopolymer is distinguished by strong fluorescence which is proportional to metal content as distinct from complexes produced by the interaction of $EuCl_3$ with macroligands containing 2-carboxybenzoyl (14) or 2-carboxynaphthoyl (15) groups (Fig. 7) [67]. A similar method was used to copolymerize MMA and MCM of Eu(III) with 1-(vinylphenyl)-3-phenyl-1,3-propandione [89].

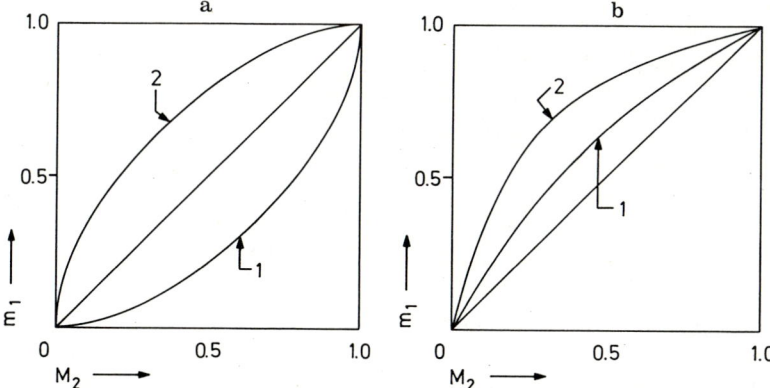

Fig. 8. Relationship between molar fractions of monomer II in copolymer (m_1) and monomer (M_2): (a) copolymerization of AN with PIP (*1*) and Co(III)-PIP (*2*); (b) copolymerization of St with PIS (*1*) and Cu(II)-PIS (*2*)

Table 6. Copolymerization activity of the MCM

Monomer I	Monomer II	Constant of monomer I copolymerization	Constant of monomer II copolymerization
AN	PIP	2.30	0.40
AN	Co(PIP)$_3$	0.86	8.60
St	PIS	0.50	1.80
St	Cu(PIS)$_2$	0.45	6.90

Radical copolymerization of MCM obtained by the reaction of Co(OCOCH$_3$)$_2$ with N,N'-bis[4'-(p-vinylbenzoyloxy)salicylaldehyde]-1,2-diaminocyclohexane with St and DVB gave a ternary product of with twenty St and DVB units per MCM unit [90].

Apart from these qualitative studies, detailed consideration has been given [91] to the copolymerization involving St, acrylonitrile (AN), MMA or acrylic acid (AA) (monomer I) and MCM (monomer II) representing complexes of Cu(II) or Co(III) with pyrrolylmethyleneiminopropene (PIP) or pyrrolylmethyleneiminostyrene (PIS). As can be seen in Fig. 8, the molar fraction of monomer II in the resulting copolymers is higher for the MCM than for its "metal-free" analogue. The reactivity of the comonomer increases after binding a metal (Table 6), which is a comparatively rare case in the polymerization and copolymerization of metal-containing monomers. Moreover, the copolymerization involves no more than one or two groups out of three coordinated with Co(III), one multiple bond remaining in the copolymer, too. In the case of a Cu(II) chelate the process involves the two vinyl groups. However, the initiation of this MCM polymerization proceeds in a complicated way:

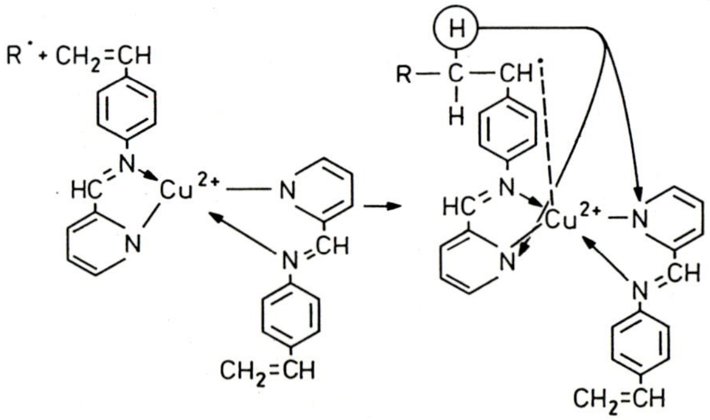

The primary radical first attacks the vinyl group of the monomer and then the free electron from this group is transferred to the central metal ion. This is stimulated by the presence of a conjugation chain (long coplanar π-electronic system) and chain propagation does not start until a complete reduction of Cu(II) to Cu(I). Simultaneously proton β-elimination from the vinyl group, which is attacked by the primary radical, and its addition to pyrrol nitrogen occur.

No quantitative data exist for the copolymerization of macrocyclic MCM. Thus attempts to copolymerize hemin with such monomers as vinylpyrrolidone (VPd) and vinylimidazole (VIm) were unsuccessful [92]. The comonomers are classed into two types, viz., π-conjugated on nonconjugated. Conjugated monomers copolymerize with macrocyclic MCM as follows:

$$\sim CH_2-\underset{R}{CH}\cdot + \underset{\underset{CH_2=CH}{|}}{\underset{/Fe^{3+}/}{CH_2=CH}} \longrightarrow \sim CH_2-\underset{R}{CH}-CH_2-\underset{\underset{CH_2=CH}{|}}{\underset{/Fe^{2+}/}{CH-}}$$

Spectroscopy suggests [92] that Fe(III) reduces to Fe(II) in the course of copolymerization. Nonconjugated monomers do not copolymerize with macrocyclic MCM because of the radical being added to the hemin to give stable adducts:

$$2CH_3-\underset{\underset{CN}{|}}{\overset{\overset{CH_3}{|}}{C}}\cdot + \underset{\underset{CH_2=CH}{|}}{\underset{/Fe^{3+}/}{CH_2=CH}} \longrightarrow CH_3-\underset{\underset{CN}{|}}{\overset{\overset{CH_3}{|}}{C}}-CH_2-\underset{\underset{CH_3-\underset{\underset{CH_3}{|}}{\overset{\overset{CN}{|}}{C}}-CH_2-CH-}{|}}{\underset{/Fe^{3+}/}{CH-}}$$

It is interesting that nonconjugated monomers can copolymerize with hemin in the presence of a conjugated (third) comonomer, e.g., resulting in the appearance of a ternary hemin — VIm — St copolymer [85, 92, 93]. Copolymerization of a heme with VPd under ^{60}Co γ-irradiation or with hydroxymethacrylate has been described [94]. Tetra(p-vinyl)benzyl esters of a dinuclear Mg(II)-chlorophyll complex copolymerize by the cationic mechanism with α-methylstyrene. Copolymerization of vinylporphyrin Co(II), Fe(II) and Ni(II) monomers, containing one or two side vinyl groups, with acrylamide or N,N'-methylenebisacrylamide gives a three-dimensional structure with a parallel immobilization of a metallomonomer within the polymeric gel [95]. The parameters of such a cross-linked structure are readily controlled by the comonomers ratio.

The soluble metalloporphyrin-containing polymers are formed by the copolymerization of MMA or 4-VP with macrocyclic MCM — an interaction product of acrylic acid chloride with tetra-p-aminophenylporphyrinate acetate manganese [96]. Copolymers obtained by the radical copolymerization of acryloyl derivatives of cobalt phthalocyanine with 9-vinylcarbazole [97] should also be mentioned.

Data on MCM graft polymerization are extremely limited. A method of preirradiation in air has been used to effect the graft polymerization of Co(II), Ni(II) and Cr(III) acrylate complexes with 2,2'-dipyridyl and 1,10-phenantroline onto a polytetrafluoroethylene powder [11f].

One of the most wide-spread approaches to the assemblage of PCMU consists in polycondensation, i.e. interaction of diketones, diamines and MX_n. Thus at 120 °C in the presence of $CoCl_2$, $NiCl_2$ or $CuCl_2$ one can obtain [98] high-molecular weight Schiff bases with metal ions in the form of macrochelates. Of great interests is the assemblage of a dimethylglyoxyme cobalt complex on a polymeric carrier [99]

$$DH_2 + Co(OCOCH_3)_2 + KCN + P4VP \rightarrow$$

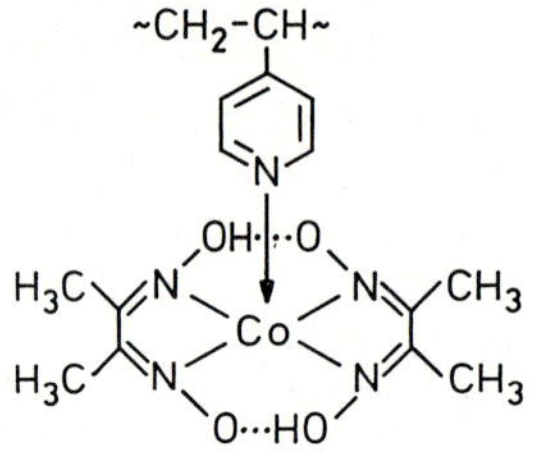

where P4VP is poly-4-vinylpyridine. Various methods of assemblage are used for the preparation of organometallic compounds on polymeric carriers. Thus a complex incorporating *trans*-bis Pd-C bonds is formed [100] in the interaction of K_2PdCl_4 with 2,6-pyridine propanoate derivatives in the pressence of P4VP or copolymer of St and 4-VP (St-*co*-4-VP) as follows:

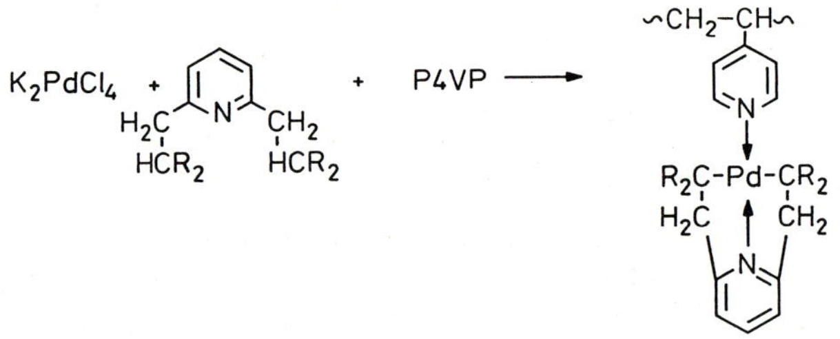

where R = $COOCH_3$ or $COCH_3$.

The method of arising reagents (*in statu nascendi*) is widely used for the preparation of polymeric phthalocyanine and porphyrins. Among the numerous possibilities we shall note only the formation of such phthalocyanines in urea, pyromellitic anhydride and anhydrous metal chloride melt in the presence of NH_3 (a polymerization degree of 5–10) [101] and also from tetracyanbenzene and NbS_2 and TaS_2 in the presence of urea [102]. Production of poly(octacyclophthalocyanine) of Cu(II) by reaction of tetracyanbenzene with $CuCl_2$ and 1-methyl-2-pyrrolidone has been described in great detail [103]. The polycondensation method involving ready-made metallochelates can be illustrated by the following interesting scheme of conversions including the condensation of a Pd(II) complex with toluene diisocyanate [104]:

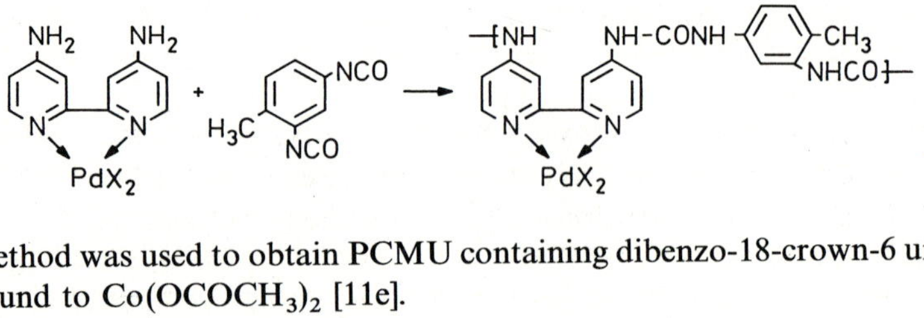

A similar method was used to obtain PCMU containing dibenzo-18-crown-6 units complex-bound to $Co(OCOCH_3)_2$ [11e].

An effective way for the preparation of PCMU is the interaction of polymers or copolymers with the LMWM occuring without destruction of the chelate node. Such an immobilization is effected through a coordination or valent binding of metal to the functional groups of polymeric carriers. The coordination metal bond to the macromolecular carrier is formed in the case of polymers containing main groups with σ-donor or π-acceptor properties. The coordination chemistry of such ligands is close to the chemistry of low-molecular weight axial ligands. The solubility and structure of PCMU are affected by the degree of coordination in the complex (a molar ratio of the chelate per monomeric unit of polymer), degree of ligand polymerization, number and type of the counterions. In particular, note that in most cases Co(III) chelates are coordinated monoaxially so that there is no intermolecular cross-linking. If, on the other hand, PCMU are produced by the method of arising reagents [105] or in stringent conditions [106], a polymeric ligand is cross-linked with Co chelates. Co(II) chelates are coordinated both mono- and diaxially whereas a low-molecular weight analogue (pyridine) gives rise to monoaxial adducts only. Formation of diaxial complexes is preferred owing to entropy gain or polymeric chelate effect.

Also, an important role in this case is played by the nature of solvent. For instance, interaction in the $CoCl_3(en)_2$ — poly(N-vinyl-2-methylimidazole) system in an $H_2O-C_2H_5OH$ mixture, as ethanol content increases from 0 to 10 wt% involves lowering the thermodynamic parameters $\Delta F^{\ne}$ from 11.1 to 10.4 kcal/mol, ΔH from 23.2 to 9.9 kcal/mol and $\Delta S^{\ne}$ from 39.3 to 11.3 e.u. [107].

In methanol 50–60 h) by reaction of cis-Ru(Dipy)$_2$Cl$_2$ with P4VP, [Ru(Dipy)$_2$L$_2$]Cl$_2$ was synthesized, and by addition of Py during of synthesis — [Ru(Dipy)$_2$LPy]Cl$_2$ [108]:

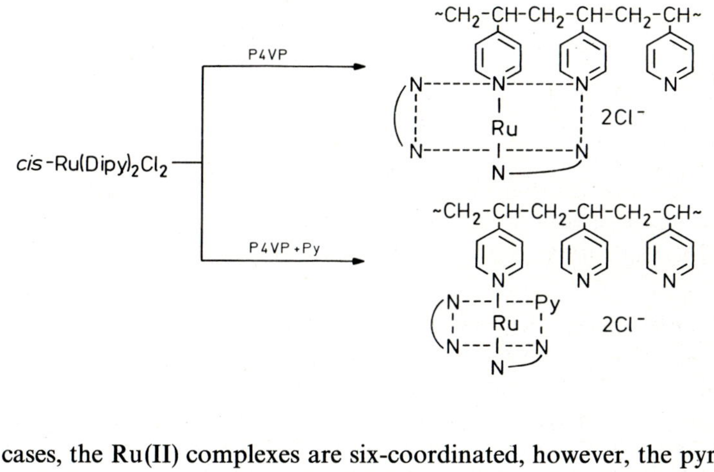

In the two cases, the Ru(II) complexes are six-coordinated, however, the pyridine complex is soluble in water whereas [Ru(Dipy)$_2$L$_2$]Cl$_2$ is insoluble because of a strong intermolecular interaction. The polymeric nature of the ligand manifests itself in a 10 nm shift of the adsorption spectrum (in the UV-region) of the macromolecular complex due to lowering its resonance stabilization, differences in the energies of the ground and d-π* excited states, and steric effects of the polymer chain. The same is true of interactions in the $[Ru(Dipy)_2]^{2+}$ — St-*co*-VP systems [11c, d].

A clearly defined ability of porphyrins and phthalocyanines to be additionally coordinated (extracoordinated) with a planar complex of ligands occupying the 5 and 6 positions in the inner coordination sphere is completely utilized during their interaction with polymeric carriers. A survey of the macromolecular ligands used for this purpose is made elsewhere [7b, e].

Monomeric metallochelates can also be immobilized on the polymer surface via a valent metal-carbon bonding. For example, the interaction of a copolymer of St and chloromethylstyrene with cobaloxime in a benzene-pyridine mixture gives PCMU with a Co—C bond [109]. N_2O_2-Metallochelates such as Co(II) complex with N,N'-bis(salicylaldehyde)ethylenediamine react with poly(chloromethylstyrene) (PCMSt) in THF at 193 K in a similar fashion [110]. The same is true of reactions of quinoline (HQ) V(V) complexes with hydroxyl-containing polymers [111]:

$$\text{Ⓟ}-OH + VOQ_2OH \rightarrow \text{Ⓟ}-OVOQ_2 + H_2O$$

According to this scheme, about 2% of the hydroxyl groups are substituted by the chelate nodes of hydroxo-bis(8-hydroxyquinolyloxovanadium). In such polymeric chelates a septa-coordination V(V) structure is assumed.

Of some interest are chelation processes occuring along with the synthesis of macromolecular ligands. This can be exemplified by a mechano-chemical activation of polyethylene terephthalate and en polycondensation to give the following complexes [112]:

$$-H_2C-H_2CO-\underset{O}{\overset{}{C}}-\underset{}{\bigcirc}-\underset{O}{\overset{}{C}}=N-CH_2-CH_2-\underset{}{\overset{H}{N}}-\underset{O}{\overset{}{C}}-\underset{}{\bigcirc}-\underset{O}{\overset{}{C}}-OCH_2-CH_2-$$

$$\frac{M}{2}$$

The principal disadvantages of the methods of PCMU assemblage consist in the occurence of various side processes, insufficient reproducibility and problems with the preparation of structurally pure metallopolymers.

3.4 Synthesis of Polynuclear Metallochelate Polymers

Polynuclear PCMU can be synthesized by means of all the above described methods for the preparation of mononuclear PCMU, such as binding a few metals with polynucleating macromolecular ligands, immobilization of the ready-made and well characterized polynuclear complexes or clusters on polymers and assemblage of polymetallic sites on polymer supports.

One of the most popular methods for the preparation of low-molecular weight polynuclear metallochelate is the interaction of MX_n with polynucleating ligands

[113]. The application of such techniques to high-molecular weight compound chemistry is complicated by problems involved in the synthesis of the corresponding macroligands. Among the examples which are few in number is the production of binuclear Co(II) and Ni(II) chelates based on PE-g-poly(N-salicyloylacrylamide) [15, 22d]:

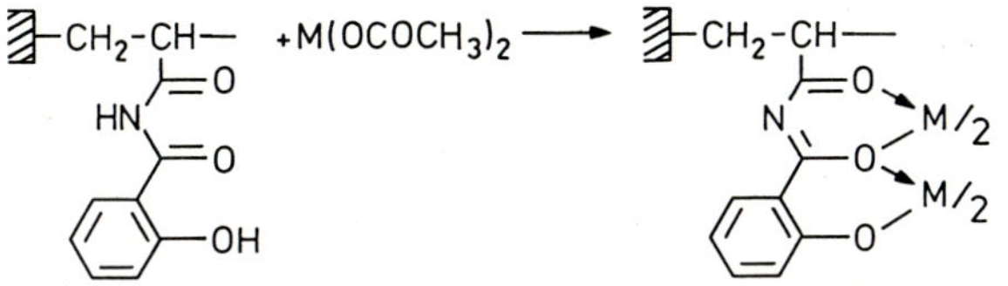

As in the case of low-molecular weight analogues [114], such systems show the presence of exchange interaction of the antiferromagnetic type between the paramagnetic centres.

Methods of stage-wise introduction of transition metals into Schiff polybases allow us to obtain heterometallic complexes with the controlled distribution of such metals (Table 7), e.g., according to the following scheme [22a, 77c]:

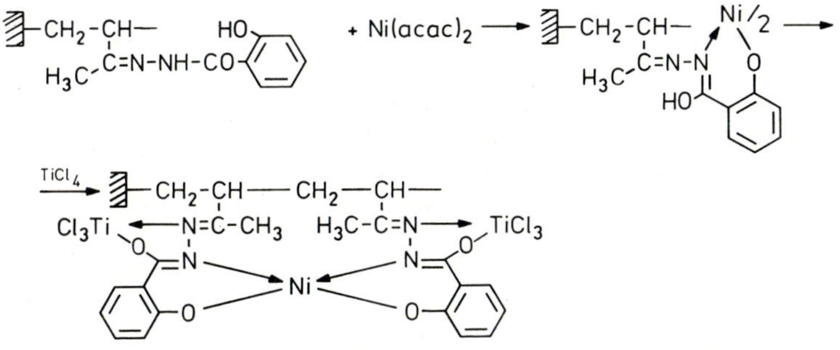

In contrast to the binding of mononuclear metallochelates with polymers, little is known about the same process for bi- and polymetallochelates. Among the most interesting examples one can cite immobilization [22b] of binuclear copper(II) complexes[4] with PCMSt:

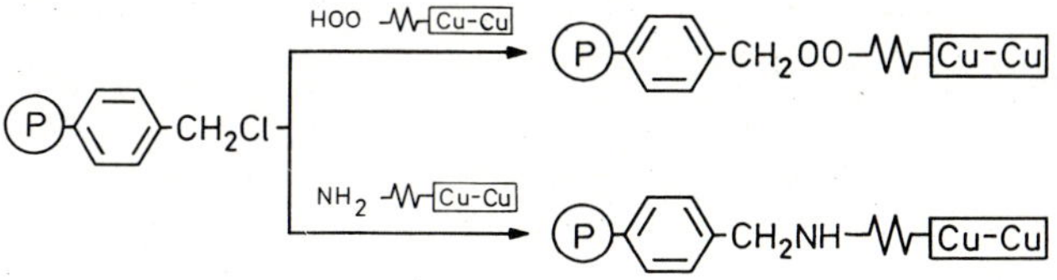

4

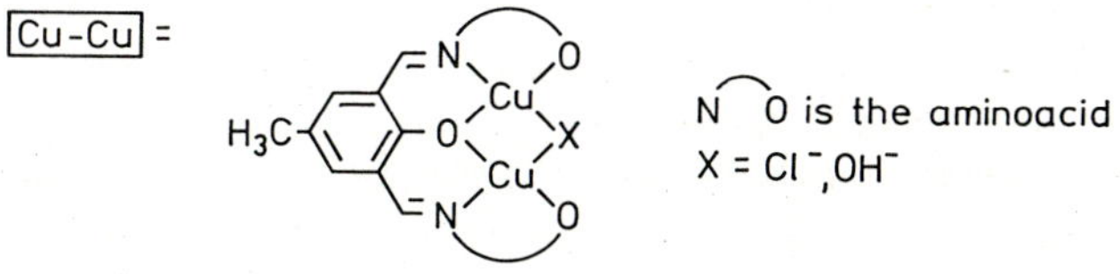

$\boxed{Cu-Cu}$ =

N͡O is the aminoacid
X = Cl$^-$, OH$^-$

Table 7. Characteristics of immobilized heterometallic complexes

Polymeric carrier	M	M'	Content of fixed metal (mmol g^{-1})		Selectivity factor (M/M')
			M	M'	
PE-g-PMVK (5 wt%) modified by salicylic acid hydrazide	Ni	Ti	0.6	0.4	1.5
	Ni	V	0.6	0.5	1.2
	Co	V	0.3	0.4	0.75
PE-g-poly (acrylic acid) (8 wt%)	Ti	Ni	0.6	0.8	0.75
	Ti	Co	0.6	0.8	0.75
	V	Ni	0.7	0.7	1.0
PE-g-poly (allyl alcohol) (4 wt%)	V	Cu	0.4	0.25	1.6
	V	Co	0.4	0.3	1.3
	V	Ni	0.4	0.3	1.3
	Ti	Cu	0.17	0.15	1.1
	Ti	Co	0.20	0.16	1.25
	Ti	Ni	0.18	0.12	1.5

or polymers containing oxirane groups:

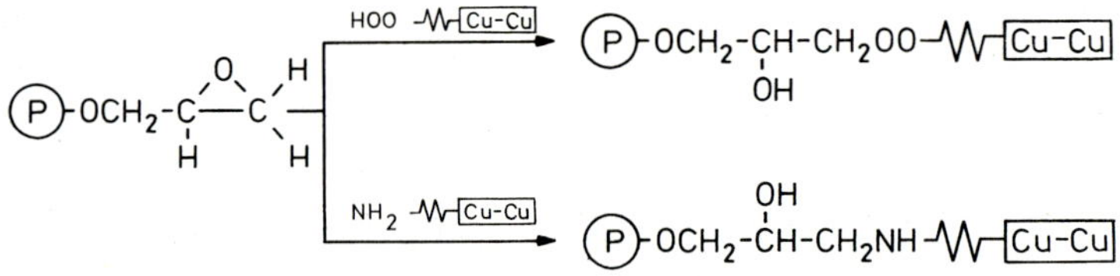

$$\text{(P)}-OCH_2-\underset{H}{\overset{H}{C}}\!\!-\!\!\underset{H}{\overset{O}{C}} \quad \xrightarrow[NH_2\ -\!\!\text{W}\!-\!\boxed{Cu\text{-}Cu}]{HOO\ -\!\!\text{W}\!-\!\boxed{Cu\text{-}Cu}} \quad \begin{array}{l} \text{(P)}-OCH_2-\underset{OH}{CH}-CH_2OO-\!\!\text{W}\!-\!\boxed{Cu\text{-}Cu} \\[1em] \text{(P)}-OCH_2-\underset{OH}{CH}-CH_2NH-\!\!\text{W}\!-\!\boxed{Cu\text{-}Cu} \end{array}$$

Diperchlorate of N,N'-trimethylene-bis(3-formyl-5-methylsalicylaldehyde iminate)-dicopper was immobilized on polyvinylamine by the following scheme [22c]:

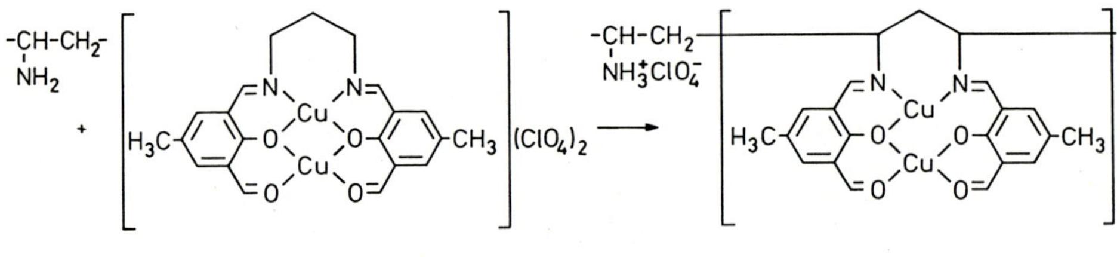

The ESR spectrum for the binuclear PCMU shows the presence of a signal with g = 2.210 whereas the low-molecular weight analogue is characterized by a broad line owing to a strong exchange interaction of the antiferromagnetic type between Cu(II) ions. Thus, one can see the disappearance of exchange interaction upon immobilization of the binuclear metallochelate on the polymer.

The fixation of cluster $Os_3(CO)_{12}$ on P4VP [115] should also be mentioned

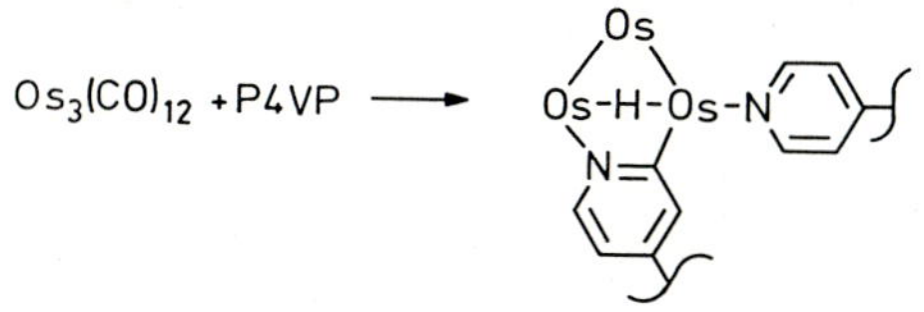

Methods suitable for the assembling of polynuclear PCMU have gained acceptance only in the last few years although they have long been fairly popular in the chemistry of LMWM [113]. Thus, a technique for preparing bimetallic complexes (Table 7) by means of the reactive outersphere (peripheral) groups of ligands or opening the coordination node of the mononuclear metallochelates has been described [22d, 77c]

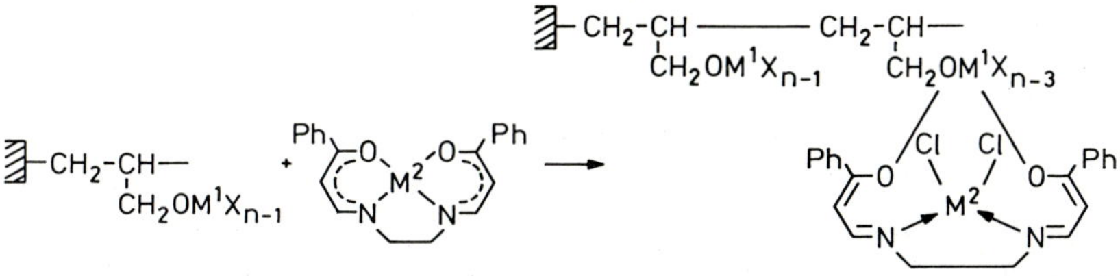

The analysis of the existing data suggest that polynuclear PCMU represent one of the least studied classes of polymeric chelates with the methods of production being in their infancy. In this respect we should pin our hopes on the method suitable for the polymerization of polynuclear MCM, the development of which is restricted by problems with their synthesis.

4 Structure of Polymers Containing Metallochelate Units

4.1 Structural Features of Metallochelate Containing Soluble Polymers

The processes of complexing with participation of macroligand solutions have been reviewed in a number of works [5, 13a, 32]. We shall only note here that due to the mobility of the polymer chain of a soluble CML one can usually observe a change in its conformation at the binding site (a change in the "local rigidity") necessitated by the electronic configuration of a transition metal. In other words, chelation brings about a buckling of the chain and renders its conformation more favourable for the reaction. Accordingly, the behaviour of chelating CML fragments is similar to that of their low-molecular weight analogs. In particular, polymers containing Dipy groups retain the complexing ability of 2,2'-dipyridyl [11g] in that they have a high affinity for transition metal ions, are less effective in the binding of non-transition metals and are absolutely incapable of complexing with alkaline metal ions.

As a rule, MX_n form two or more bonds in the chelation with their own or any other chains (intra- and intermolecular varieties, respectively). Usually, intramolecular complexes are formed in dilute solutions [116]. Thus, the interaction of poly(5-vinyl-6,6'-dimethyl-2,2'-dipyridyl) with MCl_2 (M = Co, Ni, Cu) in ethanol at 20 °C gives rise to 1:1 complexes via an M ← N bond [11b]. Increasing viscosity in DMF on dilution (Fig. 9) suggests a polyelectrolyte nature of the complexes. Data on the solubility point in the absence of intermolecular cross-linking in that the Dipy groups behave like a bidentate ligand in the side polymer chain. The same applies to the formation of vanadyl and Co(II) complexes with polyureas incorporating a Dipy group [11e].

However, the polymeric nature of the ligand induces changes in the local concentration of the functional groups, therefore a metal ion entering a region with a high local concentration of chelating fragments immediately gives rise to coordinate-saturated metallochelates and the remaining (free) groups form a small amount of coordinate-unsaturated structures. Therefore, when passing from monomeric to polymeric chelates the difference between the first and second stability constant will be lower [117]. When, on the other hand, there exist steric hindrances for a structure necessitated by the electronic configuration of a transition metal (e.g. a tetrahedral one in the case of a Zn(II) ion [117]), $\lg K_2 \ll \lg K_1$.

It should be mentioned that MX_n fixing on chelating polymers leads to the formation of rather stable complexes. As a rule, no dissociation of the chelate node proceeds by the reduction of transition metal. This can be demonstrated by two characteristic examples. The reduction of Rh(III), fixed on polyamides with Dipy groups, to Rh(I) by the action of H_2 proceeds [11h]

$$\text{\textit{f}-Dipy·Rh(III)} \xrightarrow[\text{NaOH/CH}_3\text{OH, 25°C}]{H_2} \text{\textit{f}-Dipy·Rh(I)}$$

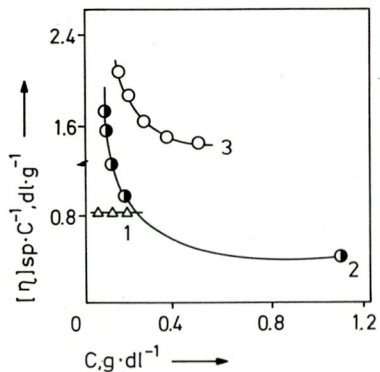

Fig. 9. Character of change of specific viscosity of poly(5-vinyl-6,6'-dimethyl-2,2'-dipyridyl) at 25 °C in DMF (*1*) and by addition of $ZnCl_2$ (molar ratio [L]/[Zn] = 0.6 (*2*) and 1 (*3*))

in two steps: at the first step Rh(I) is formed, but at the second step the autocatalytic reduction of Rh(III) with Rh(I) involved takes place according to the equation:

$$d[Rh(I)]/dt = k[Rh(III)]^2[Rh(I)]$$

The immobilized Rh(I) is stable only in a H_2 atmosphere.

Similarly, the Pd(II) reduction occurs [104]:

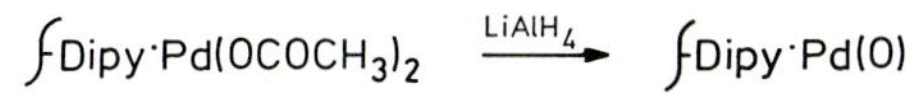

$$\int Dipy \cdot Pd(OCOCH_3)_2 \xrightarrow{LiAlH_4} \int Dipy \cdot Pd(0)$$

In the macrocomplex, Pd(O) can be oxidized again by dilute nitric acid or $(NH_4)_2Ce(NO_3)_6$.

The interaction of MX_n with CML in dilute solution involves not only changes in chain flexibility during the reaction but also bond weakening to result in extensive chain degradation. The degree of degradation increases with the number average molecular weight of the parent macroligand up to the limiting molecular weight [118]. The most probable mechanism of this process may be associated with the formation of radicals in the reaction medium, e.g. through a homolytic degradation of the metal-ligand bond in poly-β-ketoesters occuring via the radical stages as follows [13b]

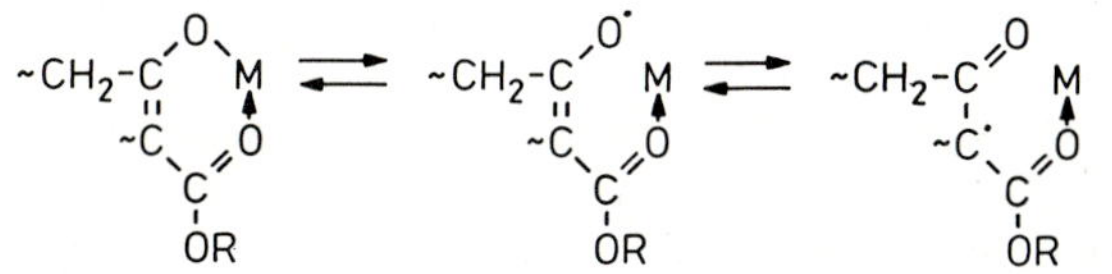

Thus analysis of the available experimental evidence suggests that soluble CML behave mostly like their low-molecular weight analogs; however, studies of such systems should be made considering their polymeric nature.

4.2 Details of Metal Binding with Cross-Linked Chelating Polymers

In the case of cross-linked macroligands their polymeric nature is responsible for a significant distortion of the geometry of the coordination sites in PCMU. For the systems under study the energy of the polymer chain "adjustment" exceeds that of chelation. Therefore, as opposed to soluble polyligands, the structure of chelate nodes is governed by the structural features of cross-linked CML. Unfortunately, data on the spatial structure of the coordination sites in PCMU are extremely scarce. For example, the most probable structure for the chelate node composed of poly(o-isophthaloylphthalamidoxime) and Cu(II) is represented by a five-membered cycle incorporating coordination bonds with C=N- and NH-groups [72]. In contrast to the planar low-molecular weight analogues, Cu(II) chelates with polymeric carriers containing salicylaldimine groups have a distorted tetrahedral structure [14d]. Depending on reaction conditions and nature of a transition metal the complexing with poly(iminoethylene)dithiocarbamate involves six, two or three thiocarbamate groups per metal atom [119]. The NCS_2 groups display their bidentate nature in binding transition metal ions whereas the thiol groups of cross-linked poly [N-(acryloylamino)methyl]mercaptoacetamide with Cu(II) react in 2:1 stoichiometry [120]. The Cu(II) ion in such a chelate node has no axial symmetry, and the most probable structure of its polyhedron is a rhombically distorted square including S_2O_2 donor groups. For the following polymeric ligands incorporating the macrocycles

16   17   18

the series of activities in metal binding (Fig. 10) [121] correspond to their low-molecular weight analogs. In this case complexes in a cavity with a metal-macrocycle ratio of 1:1 appear. However, with increasing rigidity of the immobilizing system the kinetic barrier for metal implantation becomes higher with the result that the macrocyclic ligands are not coordinated with a cavity but rather with an outer sphere [121].

It should be noted that PCMU based on cross-linked macroligands possess a relatively high chemical and thermal stability. The stability of poly(enolketonate) chelates obtained by oxidation of thin polyvinylacetate films increases in the series of mono-, di- and trivalent ions [122], the oxidation promoting the penetration of the metal ions into the deep film layers. The weight loss of Co and Mn polychelates based on the condensation products of stoichiometric amounts of 5,5′-methylene-bis-salicylaldehyde and 4,4′-diaminophenyl ether at 300, 500, and 600 °C is 2.1, and 0.5; 8.0 and 9.8; 25.0 and 27.5%, respectively [14b]. The stability, in the range between 275 and 640 °C, of chelates formed by the transition metals and condensation products of p-hydroxybenzoic acid, urea and formaldehyde follow the series [123]   Fe(III) > Co(II) > Cu(II) > Ni(II) > VO$_2$(II) > Zn(II) ≈ Mn(II) > CML.

Increasing the degree of cross-linking leads to the distortion of the chelate nodes resulting from the twisted chain strain in the formation of a second and further coordination sites [14d].

Polymer carriers with a grafted functional surface layer (i.e., the soluble functional covering on an insoluble substrate) occupies an intermediate position between soluble and cross-linked macroligands. Such polyligands present interest due to the fact that the functional groups localized on the surface facilitate the

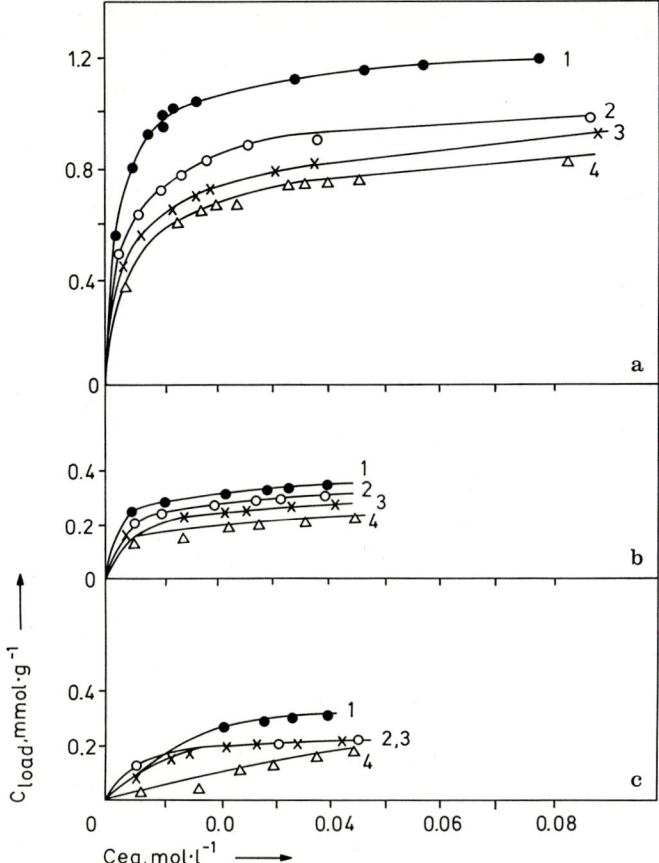

Fig. 10a–c. Plots of metal-ion loading against equilibrium concentration of copper (*1*), zinc (*2*), nickel (*3*), cobalt (*4*) nitrates for polymers *16* (**a**), *17* (**b**), *18* (**c**)

chelation process. For example, the treatment of enaminoketone-type carrier suspension with an MX_n solution gives rise to intramolecular bis-chelate complexes [15, 22d].

5 Trends in the Use of Polymers Containing Metallochelate Units

The fields of use of PCMU can be divided into two distinct groups: one is associated with the effect of metal on operating characteristics of polymers and the macromolecular chain on metal properties and the other with the development of novel polymeric materials whose properties sharply differ from those of the initial reagents. The major fields of PCMU applications are presented in Scheme 2.

5.1 Upgrading Engineering and Physico-Mechanical Polymer Properties

One principal task at the initial stage of PCMU studies consisted of the creation of thermostable polymers [2]. The studies conducted indicated that most PCMU are

Scheme 2

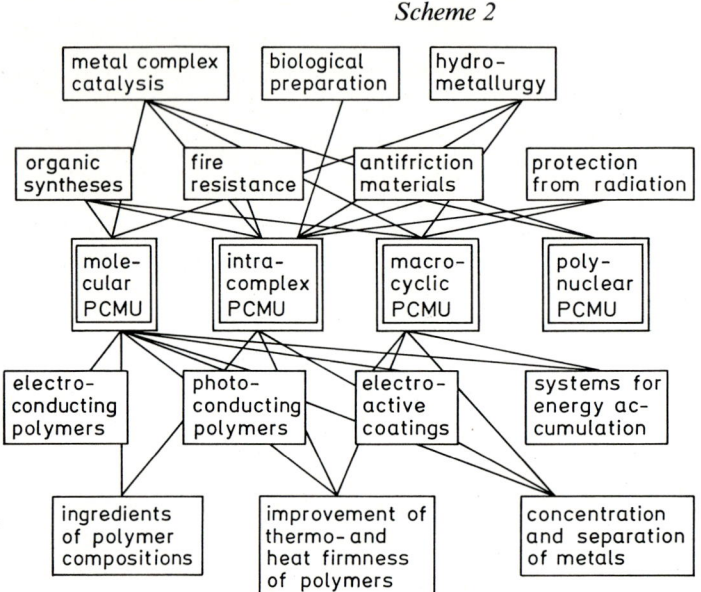

stable up to 473–523 K. Even small additions of metals to cross-linked CML enhance their stability and resistance to organic solvents. Thus compositions with such a thermal stability are obtained by immobilizing metallochelates of the β-dicarbonyl and acylhydrazone [124] types on chlorine-containing polymers. These properties are useful, especially in the case of titanium-containing polymers, for the production of various coatings, frequently with enamel-like surfaces. The thermal treatment of polymeric chelates of Ag(I), Cu(II), Zn(II), Ti(IV) with Schiff bases from β-hydroxy-γ-amines [125] at 140 °C enhances their softening point and thermostability and also imparts to them semiconductive properties due to the electronic and ionic, in the case of Ti(IV), contributions at $\sigma_{303} = 1.1 \times 10^4 - 8.0 \times 10^8$ Ohm^{-1} cm^{-1}, $E_a = 26-109$ kJ mol^{-1}.

For PCMU the σ_{298} value is mostly about 10^{-1} to 10^{-13} Ohm^{-1} cm^{-1} at $\Delta E = 0.1-1.5$ eV [126]. The presence of a combined degree of oxidation (Pd(O) and Pd(II)) is essential for providing a good conductivity of polymers alloyed with Pd compounds at the stage of polycondensation [127]. Electroactive polymeric coatings are obtained by reductive electrochemical polymerization of Fe(II), Ru(II) and Os(II) vinyl derivatives containing Dipy groups [88c].

Another extensive application of PCMU is in organic synthesis, particularly in reactions on polymer supports [128]. Among the advantages of such systems are the possibility of catalysts regeneration, automated waste-free processes, no problems with the extraction of final products and better environmental control.

Such systems can be used to advantage in the laser-microchemical industries for the direct information recording [129].

5.2 Catalytic Activity of Macromolecular Metallochelates

In the last few years, there has been a development of new branch of catalysis, i.e. the chemistry of immobilized catalysts which couple a high activity and selectivity of homogeneous- and a good workability of heterogeneous catalytic systems [5]. Chelation is one of the simplest ways for overcoming the main disadvantages of immobilized metallocomplex catalysts, i.e., a relatively low stability of the metal-polymer bond in the course of the reaction to be catalyzed. For example, a catalyst using rhodium complexes with St-co-DVB, containing Dipy fragments, for olefin hydrogenation was found to be effective for more than 300 cycles [53]. Complexes of Ni(II) with Schiff bases, immobilized on the PE surface, are active in ethylene dimerization for over 40 h whereas a homogeneous analog loses its useful activity after 2 h [22d]. Highly active catalysts for hydrogenation of alkenes, dienes, alkynes and aromatic compounds were obtained by immobilization of rhodium, palladium, ruthenium and nickel complexes with anthranilic or phenylanthranilic acid immobilized on St-co-DVB [130]. High selectivity is exhibited by catalysts obtained by interaction of $RhCl_3 \cdot 3 H_2O$ with polymers containing iminodiacetic acid groups [131]. High-active catalysts (in combination with H_2O_2) for acrylamide polymerization were obtained on the basis of an N,N'-methylene bisacrylamide — vinylporphyrin Ni(II) complex copolymer [95]. As opposed to homogeneous analogues [132], the nature of the coordination node and metallocycle of the immobilized Co(II) chelates does not greatly affect their activity and stereospecificity in 1,4-cis-polymerization of butadiene. At the same time, a considerable decrease in the specific activity, when passing from mono- to binuclear metallochelates [15] is observed for both homogeneous and immobilized systems.

At present, problems relating to immobilized complexes treated as models of polyfunctional catalysts are being extensively studied. It is assumed that such systems will allow us to realize consecutive reactions in such a way that the product of one of them will serve as a substrate for the next reaction. In optimal conditions this would make it possible to effect in a single reactor a number of conversions to obtain final products with minimum power and material expenditure [133]. This process can be exemplified by copolymerization of ethylene with butene-1 which appears on bifunctional immobilized heterometallic catalysts [22c].

5.3 Other Applications

Using PCMU as biologically active substances has been initiated only recently. A number of metallochelate polymers display fungicidal and antibacterial properties make many products resistant to attack by mold and putrefactive microorganisms [89]. The pharmacological activity of PCMU has been treated in a separate review [134]. The processes of immobilization are frequently used to control numerous reactions in the living organism but the potentialities of PCMU in this respect await further studies.

Polymer-induced chelation for analytical purposes is growing in importance [8c, d]. Among the urgent social problems associated with environmental control and power production are those relating to the extraction of toxic metals from drinking water or the production of trace amounts of metals from sea water [135]. In doing so, metal ions are selectively adsorbed, as a rule, using polymeric ligands which permit the construction of a binding node with respect to the nature of the metal ions. Thus, for example, polyallylamine modified with phosphoric acid and formaldehyde can extract up to 78% of uranium from sea water [136].

Some promise is offered by the application of PCMU to power conversion devices [137]. Thus polymeric cobalt-containing porphyrins can be used for photochemical power storage in a norbornadiene-quadricyclane system [138].

The examples cited above for PCMU applications can be enlarged. They find varied, even quite unexpected, uses. In the near future we can rely on further advances in this field.

6 Conclusion

The above analysis of the major methods for the production, structure and applications of PCMU points to a considerable progress in the solution to many problems related to PCMU chemistry and extensive studies are in progress in this field. Among the most important problems awaiting further treatment in order to disclose the general principles of PCMU chemistry we shall list the following:
1) finding new methods for PCMU production;
2) synthesis of bi- and polynuclear PCMU;
3) synthesis of mono- and polynuclear MCM together with establishing ways for the activation of their multiple bonds;
4) detailed analysis of the effect produced by the nature of metal and polymer chain on the elementary stages which form the process of chelation involving CML;
5) problems of stereoregulation and supramolecular organization of PCMU;
6) studies of the major conversions of macromolecules in the course of their chelation with metal ions;
7) examination of changes experienced by metals as they are inserted into the polymeric matrix as well as cooperative interactions between the metal ions;
8) studies of the PCMU topochemistry.

The final solution to these problems calls for the joint efforts on the part of specialists in the fields of high-molecular weight compounds, coordination chemistry, organic synthesis, catalysis, etc.

7 References

1. Pomogailo AD, Uflyand IE (1988) Koord. Khim. 14:147
2. Bailar JC (1978) In: Carraher CE Jr, Sheats JE, Pittman CU Jr (eds) Organometallic polymers. Academic, New York
3. a) Marvel CS, Tarkoy N (1957) J. Amer. Chem. Soc. 79:6000; (1958) J. Amer. Chem. Soc. 80:832
   b) Goodwin HA, Bailar JC (1961) J. Amer. Chem. Soc. 83:2467
4. Millar JR (1957) Chem. Ind. 606
5. Pomogailo AD (1988) Polymeric immobilized metal complex catalysts. Nauka, Moscow
6. Kaneko M, Yamada A (1984) Adv. Polym. Sci. 55:2
7. a) Matveeva NG (1972) In: Encyclopedia of Polymers, vol 2. Sov. Entsikl., Moscow, p 1100
   b) Tsuchida E, Nishide H (1977) Adv. Polym. Sci. 24:1
   c) Kaneko M, Tsuchida E (1981) J. Polym. Sci.: Macromol. Rev. 16:397
   d) Vinogradova SL (1980) In: Advances in the Synthesis of Organoelement Polymers. Nauka, Moscow, p 10
   e) Wöhrle D (1983) Adv. Polym. Sci. 50:45
8. a) Hering R (1967) Chelatbildende Ionenaustauscher. Akad. Verlag, Berlin
   b) Saldadze KM, Kopylova-Valova VD (1980) Complex-Forming Ionites. Khimiya, Moscow
   c) Myasoedova GV, Savvin SB (1984) Chelate-forming sorbents. Nauka, Moscow
   d) Sahni SK, Reedijk J (1984) Coord. Chem. Rev. 59:1
9. Garnovsky AD (1988) Koord. Khim. 14:579
10. a) Svec F, Kalalova E, Kalal J (1983) Angew. Makromol. Chem. 136:183
    b) Kokorin AI, Berentsveig VV, Kopylova VD, Frumkina EL (1983) Kinetics and Catalysis. 24:181
11. a) Nishide H, Shimidzu N, Tsuchida E (1982) J. Appl. Polym. Sci. 27:4161
    b) Newkome GR, Yoneda A (1983) Makromol. Chem. Rapid Commun. 4:575
    c) Kaneko M, Yamada A, Tsuchida E, Kurimura Y (1982) J. Polym. Sci.: Polym. Lett. Ed. 20:593
    d) Sumi K, Furue M, Nozakura S-I (1984) J. Polym. Sci: Polym. Chem. Ed. 22:3779
    e) Zhang K, Kumar GS, Neckers DC (1985) ibid. 23:1213, 1293
    f) Uflyand IE, Kokoreva IV, Starikov AG, Sheinker VN, Pomogailo AD. (1989) React. Polymer. 11:221
    g) Kuntz ME (1971) Bull. Soc. chim. France. 3805
    h) Wang Yu-Pei, Neckers DC (1985) React. Polym. 3:181
12. Elman B, Moberg C (1985) J. Organomet. Chem. 294:117
13. a) Davydova SL, Plate NA (1975) Coord. Chem. Rev. 16:195
    b) Davydova SL, Plate NA, Kargin VA (1968) Usp. Khim. 37:2223; (1970) Usp. Khim. 39:2256
    c) Bhaduri S, Khwaja H, Khanwalkar V (1982) J. Chem. Soc. Dalton Trans. 445
    d) Phillips HH, Kinstle JF, Adcock JL (1981) ACS Polym. Prepr. 24:320
    e) Fei CP, Chan TH (1982) Synthesis. 467
14. a) Bohlen H, Martens B, Wöhrle D (1980) Makromol. Chem. Rapid Commun. 1:753
    b) Patel MN, Patil SH (1981) Ind. J. Chem. 20A:523
    c) Patel MN, Jany BN (1985) J. Macromol. Sci. 22A:1517
    d) Kalalova E, Populova O, Stokrova S, Stopka P (1983) Collect. Czech. Chem. Commun. 48:2021
    e) Pomogailo AD, Khrisostomov FA, Djachkovsky FS (1985) Kinetics and Catalysis. 26:1104
    f) Potapov GP, Fedorova EI, Malygina OA (1982) Vysokomol. Soedin. 24B:181
15. Pomogailo AD, Uflyand IE, Golubeva ND (1985) Kinetics and Catalysis. 26:1404
16. a) Carlini C, Sbrana G (1981) J. Macromol. Sci. 16A:323
    b) Sigel H (ed) Metal ions in biological systems, vol 9. Marcel Dekker, New York
    c) Lekchiri A, Morcellet-Sauvage J, Morcellet M (1987) Macromolecules. 20:49

17. a) Patel AU, Patel HS, Patel MN (1985) Angew. Makromol. Chem. 131:135
    b) Patel BS, Lad MJ, Patel SR (1984) J. Macromol. Sci. 21A:105
    c) Warshawsky A, Kalir R (1979) J. Appl. Polym. Sci. 24:1125
18. a) Shirai H, Higaki S, Hanabusa K, Kondo Y, Hojo N (1984) J. Polym. Sci.: Polym. Chem. Ed. 22:1309
    b) Gebler M (1981) J. Inorg. Nucl. Chem. 43:2759
    c) Brouwer WM, Piet P, German AL (1984) Makromol. Chem. 185:363
19. a) Yamakita H, Hayakawa K (1980) J. Polym. Sci.: Polym. Lett. Ed. 18:529
    b) Shigehara K, Shinohara K, Sato Y, Tsuchida E (1981) Macromolecules. 14:1153
20. a) Hiraoka M (1982) Crown compounds: Their characteristics and application. Elsevier, Amsterdam
    b) Smid J (1981) Makromol. Chem. Suppl. 5. 182:203
    c) Davydova SL, Barabanov VA (1980) Koord. Khim. 6:823
21. a) Gramain P, Frere Y (1979) Macromolecules. 12:921
    b) Gramain P, Frere Y (1981) Ind. Eng. Chem. Prod. Res. Develop. 20:524
22. a) Uflyand IE, Pomogailo AD, Golubeva ND, Sheinker VN (1985) In: Proc. XVII Europ. Congr. Molecul. Spectrosc. Madrid, Spain, p P-152
    b) Prasanta S, Vigee GS (1984) Inorg. chim. acta. 90:73
    c) Koyama T, Kurose A, Masuda E, Hanabusa K, Shirai H, Hayakawa T, Hojo N (1986) Makromol. Chem. 187:763
    d) Uflyand IE, Pomogailo AD, Gorbunova MO, Starikov AG, Sheinker VN (1987) Kinetics and Catalysis. 28:613
    e) Uflyand IE, Pomogailo AD, Golubeva ND, Starikov AG (1988) Kinetics and Catalysis. 29:885
    f) Uflyand IE, Kuzharov AS, Gorbunova MO, Sheinker VN, Pomogailo AD. React. Polym. (in press.)
    g) Uflyand IE, Sheinker VN, Bulatov AV, Pomogailo AD (1989) J. Mol. Catal. 55:391
23. Cahn RS, Dermer OC (1979) Introduction to chemical nomenclature, Butterworth, London
24. Chapin EC, Twohig EF, Keys LD, Gorski KM (1982) J. Appl. Polym. Sci. 27:811
25. Kim SJ, Takizawa R (1975) Makromol. Chem. 176:891, 1217
26. Joshi RM, Patel MN (1982) Ind. J. Chem. 21A:637
27. Tolmachev VN, Gnidenko VI, Lugovaya ZA (1973) In: Paper Abstract of XV Int. Conf. Coord. Chem. Moscow, p 119
28. Polinsky AS, Pshezhetski VS, Kabanov VA (1983) Vysokomol. Soedin. 25A:72
29. Tropsha EG, Polinsky AS, Yaroslavov AA, Pshezhetski VS, Kabanov VA (1986) Vysokomol. Soedin. 28A:1373
30. Barrow GW (1974) Physical chemistry for the life sciences. McGraw Hill, New York
31. Guryanova EN, Goldstein IP, Romm IP (1973) Donor-acceptor bond, Khimiya, Moscow
32. Marinsky JA (1976) Coord. Chem. Rev. 19:125
33. Slota D, Marinsky JA (1978) ACS Polym. Prepr. 19:250
34. Davydova SL, Barabanov VA (1982) In: Problems of chemistry and applications of metal β-diketonates. Nauka, Moscow
35. Varon A, Rieman W (1964) J. Phys. Chem. 68:2716
36. Wong L, Smid J (1977) J. Amer. Chem. Soc. 99:5637
37. Iwabuchi S, Nakahira T, Fukushima Y, Saito O, Kojima K (1981) J. Polym. Sci.: Polym. Chem. Ed. 19:785
38. Yatsimirsky KB (1980) Theoret. and Experim. Chem. (UdSSR). 16:34
39. Uflyand IE, Vainstein EF, Pomogailo AD. Theoret. and Experim. Chem. (UdSSR) (in press)
40. Barbucci R, Campbell MJM, Casolaro M, Nocentini M, Reginato G, Ferruti P (1986) J. Chem Soc. Dalton Trans. 2325
41. Chung C-S (1984) J. Chem. Educ. 61:1062
42. Bolshov AA, Pomogailo AD, Leonov ID (1972) In: Complex proceeding of Mangyshlak Petroleum, vol 4, Nauka, Alma-Ata (USSR), p 110

43. Casolaro M, Nocentini M, Reginato G (1986) Polym. Commun. 27:14
44. Gergely A, Nagypal I (1977) J. Chem. Soc. Dalton Trans. 1104
45. Landau LD, Livshits EM (1965) Statistical physics. Nauka, Moscow
46. Kitaigorodsky AI (1971) Molecular crystals. Nauka, Moscow
47. Winston A, McLaughli GR (1976) J. Polym. Sci.: Polym. Chem. Ed. 14:2155
48. Kokorin AI, Kolot VN, Kudryavtsev GI, Rudaya LI, Kvitko IYa (1980) Vysokomol. Soedin. 22B:338
49. Zamarayev KI, Molin YuN, Salikhov KM (1977) Spin exchange. Nauka, Novosibirsk (USSR)
50 Kabanov NM, Kokorin AI, Rogacheva VB, Zezin AB (1979) Vysokomol. Soedin. 21A:209
51. Paton R, Watton EC, Williams LR (1974) Austral. J. Chem. 27:1185
52 Palumbo M, Cosani A, Terbojevich M, Bacchion G (1976) Biopolymers. 15:2241
53. Drago RS, Nyberg ED, El A'mma AG (1981) Inorg. Chem. 20:2461
54. Naoki H, Toshifumi H, Kunio H (1979) Polym. J. 11:139
55. Barbucci R, Ferruti P (1979) Polymer. 20:1061
55. Branca M, Marini ME, Pispisa B (1976) Biopolymers. 15:2219
57. Chong-Su Cho, Shung-Woo Kang, Jong-Woo Lim (1987) Polym. Bull. 18:447
58. Rosthauser JW, Winston A (1981) Macromolecules. 14:538
59. a) Bailar JC (ed) (1956) The chemistry of the coordination compounds. Chapman and Hall, New York
    b) Jolly WL (ed) (1970) Preparative inorganic reactions, vol 1–3. Interscience, London
    c) Skorik NA, Kumok VN (1975) Chemistry of coordination compounds. Vysshaya Shkola, Moscow
    d) Yakimov MA (1978) Fundamentals of inorganic synthesis. Leningrad University Publ., Leningrad (USSR)
60. Garnovsky AD, Osipov OA, Kuznetsova LI, Bogdashev NN (1973) Usp. Khim. 42:177
61. Holm RH, Everett GW, Chakravorty A (1966) Progr. Inorg. Chem. 7:1
62. Nose Y, Hatano M, Kambara S (1966) Makromol. Chem. 98:136
63. Shetty G (1968) Helv. chim. acta. 51:509
64. Winston A, Kirchner D (1978) Macromolecules. 11:597
65. Desaraju Padma, Winston A (1985) J. Polym. Sci.: Polym. Lett. Ed. 23:73
66. Ueba Y, Zhu KJ, Banks E, Okamoto Y (1982) J. Polym. Sci.: Polym. Chem. Ed. 20:1271
67. Joshi RM, Patel MN (1983) J. Macromol. Sci. 19A:919
68. Nozawa T, Nose Y, Hatano M, Kambara S (1968) Makromol. Chem. 112:73
69. Stepanov FA, Zosim LA (1967) Ukr. Khim. Zh. (USSR). 33:485
70. Warshawsky A, Kalir R, Patchornik A (1978) J. Org. Chem. 43:3151
71. Ferruti P, Barbucci R (1984) Adv. Polym. Sci. 58:55
72. Lehtinen A, Purokoski S, Lindberg JJ (1975) Makromol. Chem. 176:1553
73. Mitchell PCH, Taylor MG (1982) Polyhedron. 1:225
74. Holy NL, Shalvoy R (1980) J. Org. Chem. 45:1418
75. Aeissen H, Wöhrle D (1981) Makromol. Chem. 182:2961
76. Modelli A, Scagnolari F, Innorta G, Foffani A, Torroni S (1984) J. Mol. Catal. 24:361
77. a) Ritz J (1974) Makromol. Chem. 175:739
    b) Davydova SL, Barabanov VA, Dobrovolskaya NV, Plate NA (1970) Izv. Akad. Nauk SSSR. Ser. Khim. 475
    c) Pomogailo AD (1981) Metal complexes fixation on macromolecular carriers and catalytic properties of immobilized system in polymerization processes, DSc Thesis, Institute of Chemical Physics of the USSR Academy of Sciences, Moscow
78. a) Lavrentyev IP, Khidekel ML (1983) Usp. Khim. 52:596
    b) Garnovsky AD, Ryabukhin YuI, Kuzharov AS (1984) Koord. Khim. 10:1011
79. Brito H, Brito V, Springer I (1977) Makromol. Chem. 178:2507
80. Kurimura Y, Yokota H, Shigehara K, Tsuchida E (1982) Bull. Chem. Soc. Japan 55:55
81. Manecke G, Wille R (1970) Makromol. Chem. 133:61; (1972) ibid. 160:111
82. Pomogailo AD, Savostyanov VS (1988) Metal-containing monomers and polymers on their basis. Khimiya, Moscow

83. Zarnegar PP, Whitten DG (1971) J. Amer. Chem. Soc. 93:3776
84. a) Naumberg M, Duong NVK, Graudemer A (1970) J. Organomet. Chem. 25:231
    b) Schrauzer GN, Windgassen RJ (1967) J. Amer. Chem. Soc. 89:1999
85. Fuhrhop J-H, Resecke S, Vogt W, Ernst J, Subramanian J (1977) Makromol. Chem. 178:1621
86. a) Clear JM, Kelly JM, O'Konnell CM, Vos JC (1981) J. Chem. res. Part. S. 260; Part M. 3039
    b) Abruna HD, Denisevich P, Umana M, Meyer TJ, Murray RW (1981) J. Amer. Chem. Soc. 103:1
    c) Calvert JM; Schmehl RH, Sullivan BP, Fassi JS, Meyer TJ, Murray RW (1983) Inorg. Chem. 22:2151
    d) Powers MJ, Galahan RW, Salmon DJ, Meyer TJ (1976) Inorg. Chem. 15:894
    e) Ellis CD, Margerum LD, Murray RW, Meyer TJ (1983) Inorg. Chem. 22:1283
87. a) Uflyand IE, Pomogailo AD, Sheinker VN (1987) In: Proc. XXV Int. Conf. Coord. Chem. Nanjing, China, p 497
    b) Uflyand IE, Ilchenko IA, Sheinker VN, Pomogailo AD (1989) In: Radical Polymerization. Proc. All-Union Conf. Gorky, USSR, p 95
    c) Zagnij VV, Syromyatnikov VG, Sinyavsky VG (1987) Vestn. Kiev Univ. Khim. 28:37
88. Annenkova VZ, Annenkova VM, Ugryumova GS (1985) Plastmassy. 11
89. Sheats JE, Carraher CE Jr, Pittman CU Jr (eds) (1985) Metal-containing polymer systems. Plenum, New York
90. Fujii Y, Kikuchi K, Matsutani K, Ota K, Adashi M, Syoji M, Haneishi I, Kuwana Y (1984) Chem. Lett. 1487
91. Tomono T, Honda K, Tsuchida E (1974) J. Polym. Sci.: Polym. Chem. Ed. 12:1243
92. Nishide H, Shinohara K, Tsuchida E (1981) J. Polym. Sci.: Polym. Chem. Ed. 19:1109
93. Kokufuta E, Watanabe H, Nakamura I (1981) Polym. Bull. 4:603
94. Nishide H, Kato M, Tsuchida E (1981) Europ. Polym. J. 17:579
95. Potapov GP, Aliyeva MI (1983) Izv. Vyssh. Uchebn. Zaved. Khim. i Khim. Tekhnol. 26:1122
96. Solovjeva AB, Samokhvalova AI, Lebedeva TS (1986) Dokl. Akad. Nauk SSSR. 290:1383
97. Itch H, Kondo S, Masuda E (1986) Makromol. Chem. Rapid Commun. 7:585
98. Shelikh AF, Tikhomirov BI, Khlopotova IA (1975) Vysokomol. Soedin. 7B:16
99. Sasaki T, Matsunaga F (1968) Bull. Chem. Soc. Japan. 41:2440
100. Newkome GR, Yoneda A (1985) Makromol. Chem. Rapid Commun. 6:77
101. Kreja L, Blewka A (1982) Angew. Makromol. Chem. 102:45
102. Bellido J, Cardoso J, Akashi T (1981) Makromol. Chem. 182:713
103. Lin JW-P, Dudek LP (1985) J. Polym. Sci.: Polym. Chem. Ed. 23:1579, 1589
104. Zhang K, Neckers DC (1983) J. Polym. Sci.: Polym. Chem. Ed. 21:3115
105. Biedermann H, Wickmann R (1972) Z. Naturforsch. 27:1332
106. Bedetti R, Carunchio V, Chernia E (1975) J. Polym. Sci.: Polym. Lett. Ed. 13:329
107. Tsuchida E, Shigehara K, Kurimura Y (1974) J. Polym. Sci.: Polym. Chem. Ed. 12:2207
108. Simidzu T, Izaki K, Akai Y, Iyoda T (1981) Polym. J. 13:889
109. Nishikawa H, Terada E-I, Tsuchida E, Kurimura Y (1978) J. Polym. Sci.: Polym. Chem. Ed. 16:2453
110. Nishinaga H, Tomita H (1980) J. Mol. Catal. 7:179
111. Aliwi SM, Bamford CH (1977) Polymer. 18:373, 381
112. Simionescu C, Oprea CV, Neguleanu C (1973) Makromol. Chem. 163:75
113. Kogan VA, Zelentsov VV, Osipov OA, Burlov AS (1979) Usp. Khim. 48:1208
114. Uflyand IE, Ryabukhin YuI, Vysotsky BD, Askalepov VN, Kurbatov VP (1982) Koord. Khim. 8:922
115. Bhaduri S, Khwaja H, Narayanan BA (1984) J. Chem. Soc. Dalton Trans. 2327
116. Morawetz H (1963) Macromolecules in solution. Interscience, New York
117. Haruta M, Harada S (1974) Makromol. Chem. 175:2585
118. Pomogailo AD, Kuzayev AI (1981) Dokl. Akad. Nauk SSSR. 256:132
119. Patel BS, Lad MJ (1984) J. Macromol. Sci. 21A:105

120. Deratani A, Sebille B, Hommel H, Legrand AP (1983) React. Polym. 1:261
121. Paredes RS, Valera NS, Lindoy LF (1986) Austral. J. Chem. 39:1071
122. Huang SJ, Wang IF (1983) ACS Polym. Prepr. 24:320
123. Parmar JS, Patel MR, Patel MM (1981) Angew. Makromol. Chem. 93:1
124. Osima Josiro, Sato Tamotsu, Koyama Syu, Fujita Tomoaki (1984) Japan Pat. 59-217746; Japan Pat. 60-79010
125. Jaguar-Grodzinski J (1980) Makromol. Chem. 181:2441
126. Hanack M (1983) Chimia. 837:238
127. Furstch TA, Taylor LT, Fritz TW, Fortner G, Khor E (1982) J. Polym. Sci.: Polym. Chem. Ed. 20:1287
128. Hodge P, Sherrington DC (eds) (1981) Polymer-supported reactions in organic synthesis. Wiley, London
129. Ehrlich DJ, Tsao JY (1983) J. Vac. Sci. Techn. Ser. B. 1:96
130. Holy NL, Shalvoy R (1984) J. Org. Chem. 49:2626
131. Nakamura Y, Hirai H (1975) Chem. Lett. 823
132. Uflyand IE, Pomogailo AD, Askalepov VN, Kurbatov VP, Kuzayev AI, Torgova TV (1986) Koord. Khim. 12:685
133. Surkov NF, Davtyan SP, Pomogailo AD, Djachkovsky FS (1986) Kinetics and Catalysis 26:714
134. Sheats JE, Pittman CU Jr, Carraher CE Jr (1984) Chem. Brit. 20:709
135. Klein J In: IUPAC Macro '80. Int. Symp. Macromol. Pisa, 1980. Prepr., vol.2, p 108
136. Kobayashi S, Tokunoh M (1985) Macromolecules. 18:2357
137. Kitaje N (1984) High Polym. (Japan). 33:52
138. King RB, Sweet EM (1979) J. Org. Chem. 44:386

Editor: K. Dušek
Received October 4, 1989

New Aspects of Polymer Drugs

Mitsuru Akashi[a] and Kiichi Takemoto[b]
[a] Department of Applied Chemistry, Faculty of Engineering, Kagoshima University, 1-21-40 Korimoto, Kagoshima 890, Japan
[b] Department of Applied Fine Chemistry, Faculty of Engineering, Osaka University, 2-1 Yamadoka, Suita 565, Japan

This article reviews results on the studies of biomedical applications of functional polymers in which biologically active compounds are chemically or physically immobilized. The strategy for achieving of biological activity of polymers, the designs of them, polymers syntheses, and biological and biomedical examinations are discussed. As polymeric anticancer drugs polymeric prodrugs having 5-fluorouracil moieties were prepared and their in vivo activity was investigated using Ehrlich's or Sarcoma 180 ascites tumor-bearing mice. It was found that polysaccharide-coated-liposome encapsuled nucleic acid analogs show potent immunomodulating activities. This was done by monitoring the in vitro superoxide anion production from activated human neutrophils and in vivo from mouse peritoneal macrophages. Moreover, heparin and/or prostaglandin I_2-immobilized hydrogel was entrapped in poly(vinyl chloride). From the result of the activated partial thromboplastin time, inhibition of platelet aggregation, and hole blood clotting time using human blood it was seen that the drug-immobilized antithrombogenic biomaterials provide high antithrombogenicity.

1 Introduction 108
2 Polymeric Prodrugs Having 5-Fluorouracil 108
  2.1 Polymer Synthesis of Vinyl Monomers Having 5-Fluorouracil 108
  2.2 Solvolysis of *N*-Substituted 5-Fluorouracil 110
  2.3 Synthesis of Polymeric Drugs Having 5-Fluorouracil or Theophylline 111
  2.4 Drug Release by Hydrolysis of Polymeric Drugs. 116
  2.5 Antitumor Activity of Polymeric Drugs Having 5-Fluorouracil . . . 123
3 Nucleic Acid Analogs and Polyanionic Polymers as Polymer Drug . . . 127
  3.1 Polymer Synthesis 127
  3.2 Interactions 128
  3.3 Encapsulation of Poly(vinyladenine) and Poly(vinyladenine-alt, comaleic acid) in Liposome 132
  3.4 Potent Immunomodulating Activity 134
4 Drug-Immobilized Biomaterials 137
  4.1 Immobilization of Heparin and/or Prostaglandin I_2 137
  4.2 Release of Heparin 139
  4.3 Antithrombogenicity 140
5 References 144

1 Introduction

The term "Polymer Drugs" or "Polymeric Drugs" has been widely used for pharmaceutically active macromolecules, especially since the 173 rd National Meeting of the American Chemical Society in 1977 [1]. After a general concept of it was elucidated by Ringsdorf [2], many investigations have been devoted to this field. Recently, an international symposium on polymer drugs and polymeric drug carriers was held at Nagasaki, Japan in 1987. At the symposium, which was organized by Sunamoto [3], current problems were elucidated. As Sunamoto pointed out, it is necessary to draw an exact line between "polymer drugs" and "polymeric drug carriers" in order to develop the study in this field. The term "Polymer drugs" means the polymer itself shows its own pharmaceutical activity, although the corresponding monomeric species are biologically inactive. On the other hand, in case of a polymeric drug carrier, the polymer acts as a drug carrier. In other words, it can be named polymeric prodrugs or drug-immobilized polymers. Then too, drug-immobilized insoluble biomaterials should be placed under the same category, because the actions of drugs, such as the drug release from biomaterials and the activity of immobilized drugs, are quite important for understanding the characteristics of the biomaterials. Furthermore, the results of in vivo studies have become indispensable to the development of a better understanding.

In this article, as a polymeric anticancer prodrug, vinyl type polymers having 5-fluorouracil moieties are prepared to study the release of 5-fluorouracil [4–7] and their antitumor activities in vivo [7, 8]. Nucleic acid analogs and polyanionic polymers, as polymer drugs, are investigated and their abilities to complex with nucleic acids [4, 9–11] and immunomodulating activities in vitro and in vivo [11] are studied. As an example of drug-immobilized biomaterials the in vitro antithrombogenicity of polyvinyl chloride with heparin and/or prostaglandin I_2-immobilized in hydrogels is discussed using human blood [12].

2 Polymeric Prodrugs Having 5-Fluorouracil

2.1 Synthesis of Vinyl Monomers Having 5-Fluorouracil

Some of the derivatives of nucleosides and nucleic acid bases have remarkable pharmaceutical activities, and are often used as medicines. Basic studies on polymer drugs and polymeric prodrugs containing nucleic acid bases have been developed by Takemoto [13–18] and Pitha [19] with their co-workers. In the pyrimidine family, 5-fluorouracil (5-FU) and its derivatives are the most commonly used for cancer chemotherapy, particularly in Japan, and among the purins, mercaptpurin is used for leukemia while theophylline and its derivatives are administrated as diuretic drugs. Since the polymers containing pyrimidines or purins have no pharmaceutical activity, the monomeric active groups must be released from macromolecular domain in order to achieve their in vivo activity. The derivatives of 5-FU as prodrugs have also been developed in order to reduce the toxic side-effects and minimize delivery problems [20–26].

New Aspects of Polymer Drugs 109

For the preparation of *N*-2-methacryloyloxyethyl-5-fluorouracil (MAOFU), which is a typical polymerizable monomer containing 5-FU, the silylated 5-FU (1) was allowed to react with 2-bromoethyl methacrylate in acetonitrile at refluxing temperature; the decomposition of the silylated compound followed in a manner similar to that described previously [27]. MAOFU was obtained in 20% yield. When no solvent was used, the yield was extremely low, as is the reaction of silylated uracil with 2-bromoethyl methacrylate [27].

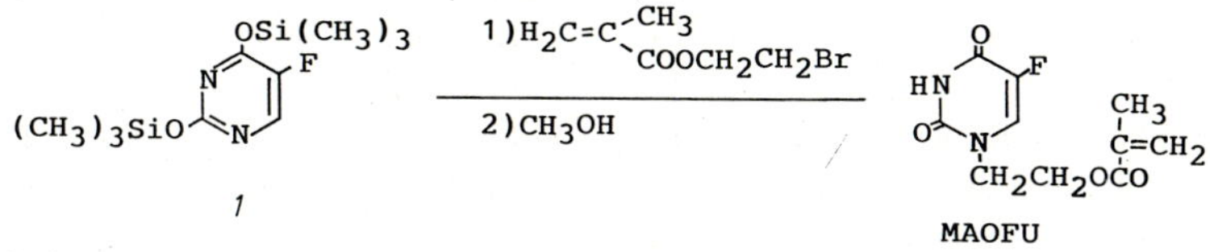

Scheme 1

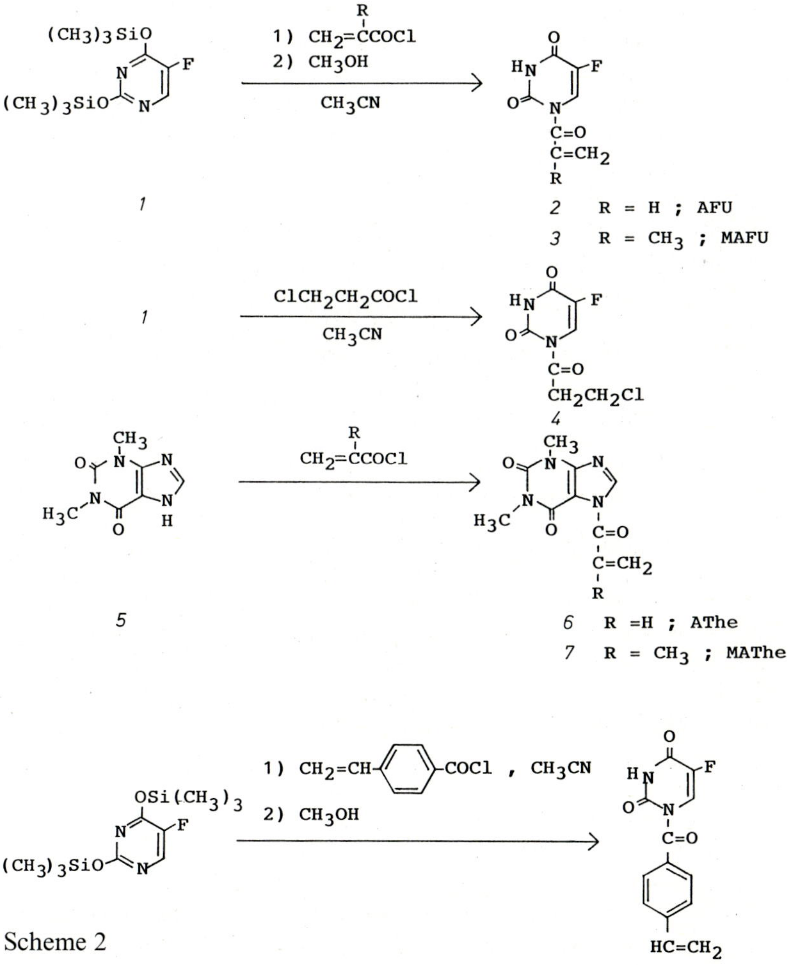

Scheme 2

When the derivatives are required to convert to the parent 5-FU in vivo, appropriate substituents were introduced across the carbonyl groups in the chemical modification [20–25]. Though the first synthesis of acryloyl derivatives of 5-FU, which is the simplest polymerizable derivative, was done by Gebelein, the monomer has not been purified and collected [23]. In the present case, as shown in Scheme 2, silylated 5-FU was used instead of just 5-FU so as to give selectivity to the 1N-acylation similar to that of the acryloyl derivatives of thymine [9]. For the preparation of acryloyl-5-FU (AFU), methacryloyl-5-FU (MAFU) and p-vinylbenzoyl-5FU (VBFU), trimethylsilylated 5-FU (1) was allowed to react with acryloyl chloride, methacryloyl chloride and vinyl-benzoyl-chloride, respectively. The reaction was carried out in water-free acetonitrile solution; after the addition of acid chloride at room temperature the solution was stirred for 30 min. This procedure afforded AFU in 16%, MAFU in 56%, and VBFU in 63% yield.

In the case of theophylline, the reaction of acryloyl chloride or methacryloyl with theophylline in dry benzene gave acryloyltheophylline (AThe) or methacryloyltheophylline (MAThe) at room temperature after 12 h in the presence of triethylamine in 75% and 65% yield, respectively.

2.2 Solvolysis of N-substituted 5-fluorouracil

As pointed out, 5-FU derivatives are prodrugs and generally have to be converted into free 5-FU after administration for pharmaceutical activity. Therefore, 1N-carbamoyl and 1N-carbonyl substituted 5-FUs have been synthesized. Scheme 3 shows the structure of 5-FU derivatives. The chemical shifts of C_6—H of pyrimidine ring of 5-FU derivatives clearly differ from 5-FU. The solvolysis rates of 1N-substituted 5-FUs, MAFU, VBFU, 1-N-stearoyl-5-FU (STFU) and 1-N-benzoyl-5-FU (BzFU) were estimated in DMSO-d_6/CD_3OD (2/1, v/v) at 60 °C by ^{1}H NMR spectroscopy. From a typical time-course of an NMR spectrum of MAFU the first-order plots are obtained (Fig. 1). The observed

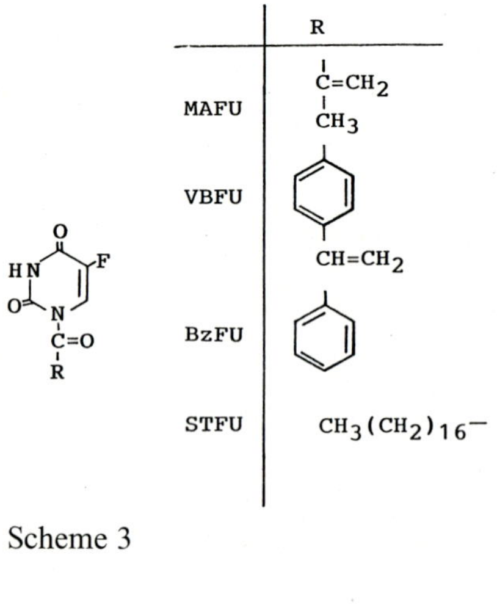

Scheme 3

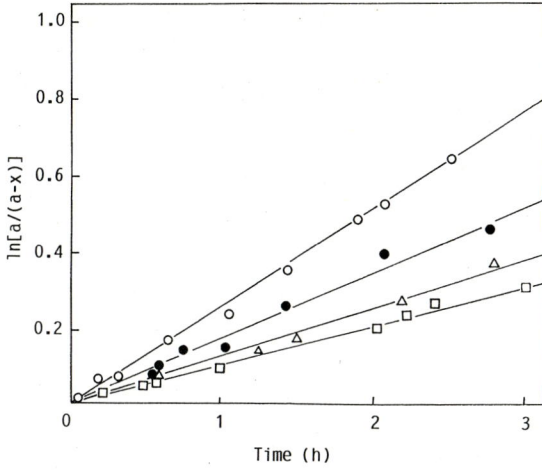

Fig. 1. Methanolysis of 5-FU derivatives in DMSO-d_6/CD$_3$OD (2/1, v/v) at 60 °C: (○) MAFU, (●) VBFU, (△) BzFU, (□) STFU

solvolysis rate constant (k) was obtained from the slope. The values of k and the half-life periods are listed in Table 1. VBFU was more stable than MAFU against methanolysis. The reactivity of STFU with lipophilic substituent at the 1N position was less than VBFU and MAFU. The same tendency has also been observed for the hydrolysis of N-acyl-5-FU [25] and N-alkoxycarbonyl-5-FU derivatives [28].

Table 1. Half-life (min) of solvolysis of 1-N-substituted 5-FU derivatives in DMSO/CD$_3$OD (2/1, v/v) at 60 °C[a]

	MAFU	STFU	BzFU	VBFU
Half-life Period	165	392	325	240
k[b] ($\times 10^3$)	4.19	1.77	2.13	2.89

[a] Concentration, 0.6 M.
[b] Rate constant (1/min)

2.3 Synthesis of Polymeric Drugs Having 5-FU or Theophylline

MAOFU was polymerized in the presence of AIBN, and the results are shown in Table 2. The polymerizability of MAOFU was almost comparable to that of the other methacryloyloxyethyl-type monomers containing thymine (MAOT) and adenine (MAOA) [29–31]. The resulting polymer, polyMAOFU, was colorless, amorphous, and soluble in DMSO, DMF, and pyridine.

Most of the polymers containing purine and pyrimidine bases are insoluble in water due to the hydrophobic nucleobase moiety [19]. In the examination of the

Table 2. Polymerization of MAOFU, MAOT, and MAOA

Monomer	Solvent	Conversion, %[a]
MAOFU	Ethanol	73
	Dioxane	56
	DMSO	72
	DMF	52
MAOT	Ethanol	76
MAOA	Ethanol	76
	Dioxane	54

[a] [Monomer] = 40 mmol/L, [AIBN] × 2 mmol/L, 10 mL of solvent, 60 °C, 6 h

biological and biomedical characters of the polymers, the hydrophilic nature of the polymers containing 5-FU could be significant. When hydrophilic polymers containing methacryloyloxyethyl type monomers are to be prepared, copolymerization with water-soluble vinyl monomers is one possible approach. In the present case, MAOFU as well as MAOA, MAOT, and acryloyloxyethyl-type monomers containing thymine (AOT) were copolymerized with acrylamide (AAm),

Table 3. Copolymerization of MAOFU, MAOT, AOT, and MAOA with water-soluble-vinyl monomers

Sample no.	M Monomer, mmol/L			Solvent	Conversion, %[a]	Content, %[b]		
	M_1	M_2	M_3			M_1	M_2	M_3
1	MAOFU (1)	AAm (19)	—	DMF	77	6	94	—
2	MAOFU (2)	AAm (18)	—	DMF	73	12	88	—
3	MAOFU (1)	AA (19)	—	MeOH	27	14	86	—
4	MAOFU (1)	MA (19)	—	MeOH	41	7	93	—
5	MAOFU (2)	MA (18)	—	MeOH	45	15	85	—
6	MAOFU (3)	MA (17)	—	MeOH	49	21	79	—
7	MAOFU (2)	AA (14)	VIm (4)	MeOH	27	12	57	31
8	MAOT (1)	AA (19)	—	MeOH	43	12	88	—
9	MAOT (1)	AAm (19)	—	DMF	68	7	93	—
10	MAOT (6)	AA (14)	—	MeOH	66	43	57	—
11	MAOT (2)	AA (9)	VIm (9)	MeOH	25	8	52	40
12	AOT (10)	AA (10)	—	MeOH	48	35	65	—
13	AOT (2)	AA (9)	VIm (9)	MeOH	19	4	56	40
14	AOT (4)	AA (12)	VIm (4)	MeOH	27	11	56	33
15	MAOA (2)	AA (18)	—	MeOH	43	23	77	—
16	MAOA (1)	AAm (19)	—	DMF	56	8	92	—
17	MAOA (2)	AA (9)	VIm (9)	MeOH	26	5	52	43
18	MAOA (4)	AA (12)	VIm (4)	MeOH	31	18	58	24
19	MAOA (1)	VP (99)	—	DMF	12	16	84	—

[a] [AIBN] = 1 mmol/L, 60 °C, 6 h.
[b] Determined by nitrogen analysis and UV

acrylic acid (AA), methacrylic acid (MA) and vinylpyrrolidone (VP) in the presence of AIBN at 60 °C. The results are summarized in Table 3. The copolymers obtained, except for the copolymer containing VP units, had sufficient water solubility for the present experiments. Electro-negative monomers, such as AA and MA, were more effective as solubilizers than electroneutral ones, such as AAm and VP.

Instead of water soluble polymeric drugs, water dispersible ones are also adequate for use in biological examination and chemotherapy. Graft copolymers having a hydrophobic backbone and hydrophilic branches have been found to have a water dispersibility [6, 32–36]. Particularly when a macromonomer is used for the graft copolymer synthesis, spherical water dispersible polymers, i.e. microspheres, are obtained in a solution [6, 33, 35, 36]. Water dispersible polymeric drugs having 5-FU or theophylline were prepared using water soluble macromonomers. A water soluble macromonomer, vinylphenyl-terminated oligovinylpyrrolidone macromonomer (St-OligoVP) ($\bar{M}_n$ = 920 and 3100) and methacryloyloxyethyl-type monomers containing theophylline (MAOThe) and MAOFU were used. Copolymerization of St-OligoVP with MAOThe or MAOFU was carried out in the presence of BPO as an initiator at 60 °C for 6 h in methanol after degassing in a sealed tube. In the course of polymerization, the graft copolymer produced was dispersed in a solution, similar to the case of copolymerization of styrene with St-OligoVP [33, 35]. After polymerization, 10 mL of acetone were added to the reaction mixture to centrifuge at 3000 rpm for 3—10 h. The graft copolymers obtained were washed, centrifuged in methanol-acetone (1/1, v/v) mixture twice, and finally dispersed in distilled water and

Table 4. Copolymerization of MAOThe (M_1) with AA or St-OligoVP (M_2) at 60 °C in methanol

Entry no.	M_1 (× 10² mol/L)	M_2 (× 10² mol/L)			Yield (wt%)	M_1 in copolymers (mol%)
	MAOThe	AA	St-OligoVP ($\bar{M}_n$ = 920)	St-OligoVP ($\bar{M}_n$ = 3100)		
1	2.0	8.0			14.1[a]	22
2	3.0	7.0			0.3	52
3	5.0	5.0			4.7	85
4	7.0	3.0			13.4	83
5	8.0	2.0			54.4	94
6	9.0	1.0			64.3	99
7	9.5	0.5			74.4	97
8	5.0		5.0		0.9	95
9	7.0		3.0		9.7	97
10	8.0		2.0		14.7	98
11	9.0		1.0		34.9	99
12	9.5		0.5		50.4	99
13	10.0				84.7[a]	100

[BPO] = 2.0×10^{-3} mol/L.
[a] Precipitated in methanol

Table 5. Copolymerization of MAOFU (M_1) with AA or St-OligoVP (M_2) at 60 °C in methanol

Entry no.	M_1 ($\times 10^2$ mol/L) MAOFU	M_2 ($\times 10^2$ mol/L) AA	M_2 ($\times 10^2$ mol/L) St-OligoVP ($M_n = 920$)	Yield (wt%)	M_1 copolymers (mol%)
14	0.5	9.5		9.3[a]	22
15	3.0		7.0	0.02	—
16	5.0		5.0	0.62	79
17	7.0		3.0	13.1	94
18	9.0		1.0	53.7	99.8
19	10.0			73.0[a]	100

[BPO] = 2.0×10^{-3} mol/L.
[a] Precipitated in methanol

freeze dried. The results of copolymerization MAOThe or MAOFU with AA or St-Oligo VP are shown in Tables 4 and 5. When a water soluble macromonomers was rich in feed, it produced water-soluble graft copolymers which were however, not collected due to their solubility under the present treatment. When hydrophobic monomers were rich in feed, they produced polymers which were found to be well dispersed in water and methanol. The molar unit of macromomomer in polymers obtained was less than that in feed, which suggests that the reactivity of St-Oligo VP is less than that of MAOThe and MAOFU. This tendency was more remarkable when using the high molecular weight macromonomer in the copolymerization of St-OligoVP with MAQThe, similar to that of St-OligoVP with styrene [33, 35]. From the scanning electron micrographs of graft copolymers having theophylline or 5-FU and OligoVP branches, spherical particles with 0.52 and 0.18 μm average diameters were found in the poly(MAOThe-co-St-Oligo VP) (Entry no. 7 in Table 4) and poly(MAOFU-co-St-Oligo VP) (Entry no. 18), respectively, while huge amorphous particles were formed in homopolymerization of MAOThe and MAOFU under the present condition.

Polymerization and copolymerization of the acryloyl or methacryloyl type monomers having 5-FU or theophylline were performed in methanol or benzene solution at 23 °C or 60 °C using AIBN as an initiator. The results of the polymerization are shown in Tables 6 and 7. The homopolymers of AFU,

Table 6. Polymerization of AFU and MAFU in methanol at 23 °C

Run	M_1	M_2	M_1/M_2 (mol%)	Conversion (%)	M_1 in Copolymer (%)
1	AFU	AAm	25/75	12	25
2	AFU	AA	25/75	41	49
3	MAFU	—	100/0	7	100
4	MAFU	AAm	25/75	24	30

[Total Monomer] = 2×10^{-1} mol/L [AIBN] = 5×10^{-1} mol/L

Table 7. Polymerization of AThe and MAThe in benzene at 60 °C

Run	M_1	M_2	M_1/M_2 (mol%)	Conversion (%)	M_1 in Copolymer (%)
5	AThe	—	100/0	16	100
6	MAThe	—	100/0	15	100
7	MAThe	MA	5/95	94	3
8	MAThe	MA	10/90	33	12

Run 5 and Run 6: [Total Monomer] = 1.0×10^{-1} mol/L. [AIBN] = 2.0×10^{-3} mol/L: Run 7 and Run 8: [Total Monomer] = 2.0×10^{-1} mol/L. [AIBN] = 2.0×10^{-3} mol/L

MAFU, AThe and MAThe were insolule in water but soluble in organic solvents such as DMSO for poly(AFU) and poly(MAFU), and dioxane for poly(AThe) and poly(MAThe). The copolymerization of water-soluble vinyl monomers such as AA were found to give water-soluble copolymers which were adequate for biological examination. As AFU and MAFU are easily hydrolyzed even by atmospheric moisture, the polymerization of AFU and MAFU was carried out in dry methanol at 23 °C to prevent the hydrolysis of monomers in the course of polymerization. The polymerizability of acryloyl and methacryloyl type monomers of 5-FU was found to be similar to that of thymine [9]. In general, the polymerizability of methacryloyloxyethyl type monomers containing purine or pyrimidine bases [29–31] is much greater than that of acryloyl or methacryloyl monomers.

Listed in Table 8 are the results of the polymerization and copolymerizations of VBFU. VBFU was quantitatively polymerized and/or copolymerized with maleic anhydride (MAn) or methacryloyl terminated poly(ethylene glycol) macronomer (PEG) to give a polymeric drug containing 5-FU. The latter afforded a

Table 8. Polymerization and copolymerization of VBFU (M_1) with M_2 by AIBN at 60 °C[a]

Run	M_2	$\dfrac{M_2}{M_1}$	Solvent	Time (h)	Yield (%)	M_1 (mol%) in polymer	Mn^b (10^{-3})	$\dfrac{Mw^b}{Mn}$
1	none	0	CH_3CN	24	80	100		
2[c]	none	0	benzene	24	90	100	[d]	
3	MAn	1	benzene	6	58	52	3.2	3.1
4	MAn	2	benzene	24	67	53		
5[e]	PEG[f]	1	CH_3CN	24	52	29	3.6	2.6
6[e]	PEG[g]	1	CH_3CN	24	52	31		

[a] $[M_1 + M_2]/[AIBN] = 50$.
[b] Estimated by GPC.
[c] $[M_1 + M_2]/[AIBN] = 100$.
[d] $\eta_{re} = 0.60$ (in DMSO at 40 °C, Concentration = 0.23 g/dL).
[e] BPO was used; $[M_1 + M_2]/[BPO] = 50$.
[f] $CH_2 = C(CH_3)-CO(OCH_2CH_2)_n-OCH_3$ (Mn = 400).
[g] Mn = 1100

water dispersible copolymer. The polymerizability of VBFU was much greater than that of methacryloyl type 5-FU monomer such as MAFU and AFU. When the concentration of VBFU was lower than that of MAn in the copolymerization, the prepared copolymers consisted of VBFU and MAn in the same molar ratio regardless of their molar ratio in feed. This is similar to the alternative free radical copolymerization of styrene and MAn or 4-vinylphenyl acetate and MAn [37]. Therefore, the present copolymer is expected to have an alternating structure. The poly(VBFU) and copoly(VBFU-co-MAn) were not soluble in water, acetonitrile, acetone, $CHCl_3$, THF, or benzene but slightly soluble in DMSO. The molecular weight of the polymer was estimated by GPC or evaluated by viscosity measurement in water after hydrolysis. The $\bar{M}_n$ of poly(VBFU-co-MAn) and poly(VBFU-co-PEG) was low (ca. 3000).

2.4 Drug Release by Hydrolysis of Polymeric Drugs

Homopolymers of methacryloyloxyethyl-type monomers containing 5-FU, thymine, or adenine were very stable toward acids and bases, while their hydrophilic copolymers were found to be degraded in aqueous solution under mild conditions to give the derivatives of purine and pyrimidines. The derivatives released from the copolymers were identified by UV spectroscopy as N-hydroxyethyl derivatives of 5-FU, thymine, or adenine after isolation by preparative TLC. This means that the ester groups of the polymer side chains were hydrolyzed in aqueous solution.

Figure 2 shows the degree of hydrolysis of poly(MAOT-co-AA) (Sample no. 10 in Table 3), poly(AOT-co-AA) (No. 15), and poly(MAOA-co-AA) (No. 12) at 60 °C in a 0.1 M phosphate buffer solution (pH 7.8) as a function of time. Acryloyloxyethyltype copolymer (poly(AOT-co-AA)) was hydrolyzed rather easily, but no significant difference between the kinds of leaving group was observed. The flexibility of the polymer chain is supposed to be an important factor for the hydrolysis of the polymer side chain.

The hydrolysis of copolymers containing vinyl imidazole (VIm) as the third component, poly(MAOT-co-AA-co-VIm) (No. 15), poly(AOT-co-AA-co-VIm) (No. 13), and poly(MAOA-co-AA-co-VIm) (No. 17), is shown in Fig. 3. These copoly-

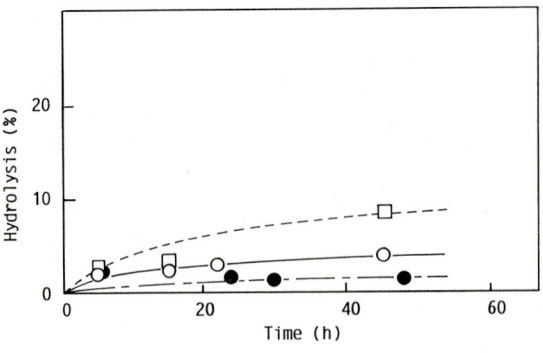

Fig. 2. Hydrolysis of poly(MAOT-co-AA) (○), poly(AOT-co-AA) (□), and poly(MAOA-co-AA) (●) at 60 °C in a 0.1 M phosphate buffer (pH 7.8). Concentration: 0.8 mg/mL

New Aspects of Polymer Drugs 117

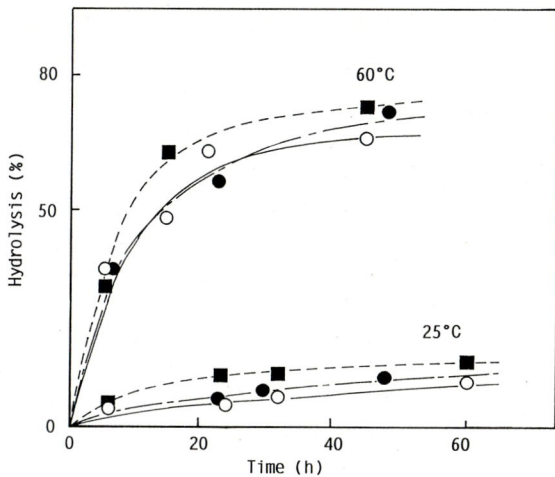

Fig. 3. Hydrolysis of poly(MAOT-co-AA-co-VIm) (●), poly(AOT-co-AA-co-VIm) (■), and poly-(MAOA-co-AA-co-VIm) (○) at 60 °C and 25 °C in a 0.1 M phosphate buffer (pH 7.8). Concentration: 0.8 mg/mL

mers were easily degraded to give the hydroxyethyl derivatives, in contrast to the hydrophilic copolymers consisting of AA and methacryloyloxyethyl type monomers. It can be assumed that the imidazole group of the polymer side chain plays a role in the ester hydrolysis as seen in the enzyme model system containing poly(VIm) [38].

To elucidate the role of the imidazole group, the concentration dependence of the hydrolysis of poly(AOT-*co*-AA-*co*-VIm) (No. 14) and poly(MAOA-*co*-AA-*co*-VIm) (No. 18) was studied and the results are shown in Fig. 4. Evidently the hydrolysis does not depend on the copolymers concentration. Therefore, it can be concluded that the ester is hydrolyzed intramolecularly. The fact that the

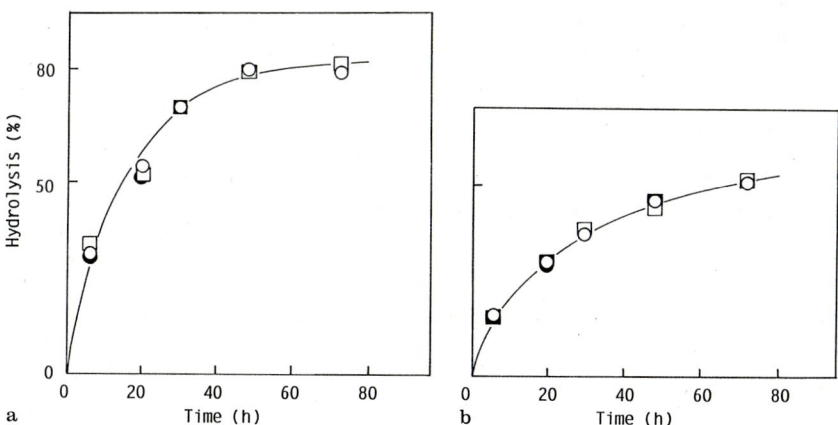

Fig. 4a, b. Concentration dependence of the hydrolysis of poly(AOT-co-AA-co-VIm) and poly-(MAOA-co-AA-co-VIm) at 60 °C in a 0.1 M phosphate buffer (pH 7.8). (a) poly(AOT-co-AA-co-VIm), (b) poly(MAOA-co-AA-co-VIm). Concentration: (○) 0.8 mg/mL, (□) 0.4 mg/mL, (●) 0.2 mg/mL, (■) 0.1 mg/mL

VIm unit of the polymer side chain does not catalyze the hydrolysis of the ester of another polymer side chain [37] also supports intramolecular catalysis by the neighboring imidazole group via six-membered ring formation as shown in Scheme 4.

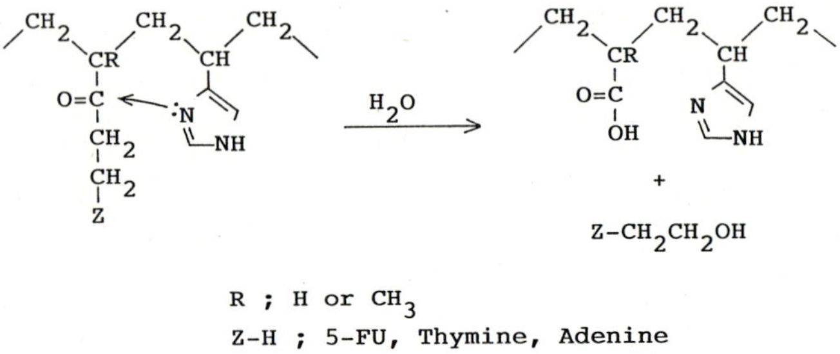

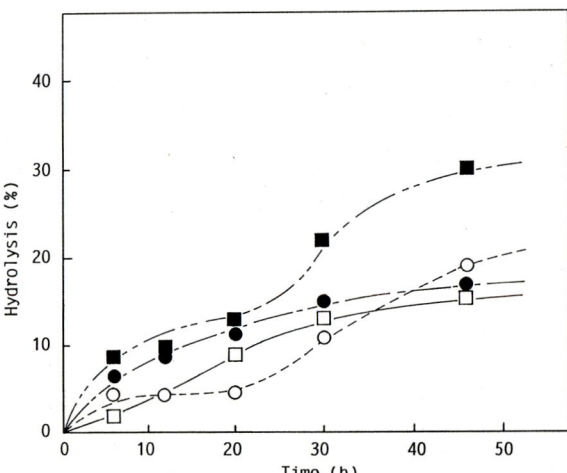

R ; H or CH₃
Z-H ; 5-FU, Thymine, Adenine

Scheme 4

Figure 5 shows the pH dependence of the hydrolysis of poly(MAOA-*co*-AA-*co*-VIm) (No. 18) at 37 °C. As shown in this figure, hydrolysis proceeds more efficiently at higher pH. This is considered to be due to the increase of the nucleophilicity of the imidazole group as a hydrolysis catalyst at higher pH.

Fig. 5. pH dependence of the hydrolysis of poly(MAOA-co-AA-co-VIm) at 37 °C. Concentration: 0.8 mg/mL. (○) pH 7.4 in a 0.1 M phosphate buffer, (□) pH 7.8 in a 0.1 M phosphate buffer, (●) pH 8.5 in a 0.05 M boric acid buffer, (■) pH 9.0 in a 0.05 M boric acid buffer

Figure 6 shows the results of hydrolysis of hydrophilic copolymers having 5-FU, poly(MAOFU-*co*-AAm) (No. 12), poly(MAOFU-*co*-AA) (No. 3), poly-(MAOFU-*co*-MA) (No. 4), poly(MAOFU-*co*-MA) (No. 5), poly(MAOFU-*co*-MA) (No. 6), and poly(MAOFU-*co*-AA-*co*-VIm) (No. 7). As shown in the figures, the hydrolysis is dependent on the kind of solubilizer, its concentration, and the temperature. The hydrolysis of the copolymer containing AAm units took place very rapidly. The reason could be because the hydroxyl anion can approach the ester group of the electroneutral copolymer rather easily. In contrast, the

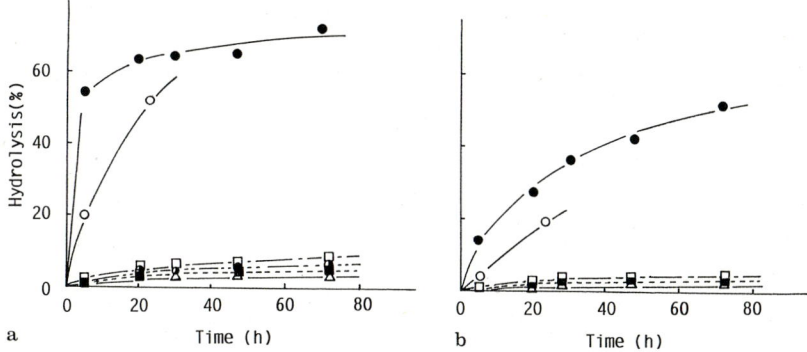

Fig. 6. Hydrolysis of copolymers having 5-FU in a 0.1 M phosphate buffer (pH 7.8). Concentration: 0.8 mg/mL. (a) At 60 °C, (b) at 37 °C. (●) poly(MAOFU-co-AAm), (□) poly(MAOFU-co-AA), (□) poly(MAOFU-co-MA), (○) poly(MAOFU-co-MA), (■) poly-(MAOFU-co-MA), (○) poly(MAOFU-co-AA-co-VIm)

electronegative charges of the carboxyl groups in the AA and MA units cause both intermolecular and intramolecular repulsion, and the approach of the hydroxyl anion to the ester group could be prevented. Furthermore, increasing content of the solubilizer and/or rising temperature could facilitate the motions of the polymer chain and thus enhance the hydrolysis. When a VIm unit was used as the third component in the copolymers, the hydrolysis proceeded readily, similar to that of hydrophilic copolymers with adenine and thymine units.

Water dispersible microspheres containing 5-FU or theophylline units release pharmaceutically active molecules as well as water soluble ones. Figure 7 shows how hydroxyethyl-5-FU and hydroxyethyltheophylline release by ester hydrolysis of graft copolymers in aqueous dispersion solution by comparing that of the water soluble copolymers in homogeneous aqueous solution. As shown in the figure, the hydrolysis depends on the chemical structure of polymeric

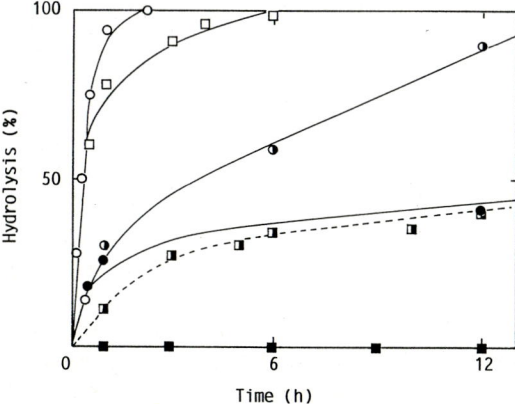

Fig. 7. Hydrolysis of polymeric drugs having theophylline or 5-FU at 37 °C. (□) poly(MAOThe-co-AA) in 0.1 M glycine buffer (pH 12.0), (□) poly-(MAOThe-co-St-OligoVP) in 5N NaOH, (■) poly(MAOThe), (○) poly(MAOFU-co-AA) in 0.1 M glycine buffer (pH 12.0), (○) poly-(MAOFU-co-St-OligoVP) in 0.1 M Glycine buffer (pH 12.0), (●) poly-(MAOFU) in 0.1 M glycine buffer (pH 12.0). Concentration: 1.0 mg/mL

drugs. It was found that the order of hydrolysis was as follows: poly(MAOFU-*co*-AA) > poly(MAOFU-*co*-St-OligoVP) > poly(MAOFU) for polymeric drugs having 5-FU, and poly(MAOThe-*co*-AA) > poly(MAOThe-*co*-St-OligoVP) = poly(MAOThe) at 37 °C in 0.1 M glycine buffer (pH 12.0), and poly-(MAOThe-*co*-St-OligoVP) > poly(MAOThe) at 37 °C in 5N NaOH solution for polymeric drugs having theophylline unit. Homopolymers of MAOThe and MAOFU were stable against hydrolysis in aqueous media because of their hydrophilicity, while their water dispersed graft copolymers were slowly degraded to give the hydroxyethyl derivatives of 5-FU and theophylline, as well as hydrophilic copolymers. Furthermore, the hydrophobicity of theophylline was usually hard to hydrolyze compared to those of 5-FU. It is concluded that since the purpose of polymeric drugs is the achievement of slow-release, when taking water soluble macromonomers for polymeric drugs containing hydrophobic drugs the resulting graft copolymers can still possess a considerable amount of pharmaceutically active groups which are hydrolyzed slowly under mild conditions. From the viewpoint of the drug delivery system, water dispersed polymeric drugs are supposed to have an advantage in medical treatment over insoluble ones. Moreover, the microspheres obtained could be directly delivered inside a living cell by phagocytosis [40].

The pharmaceutical activity of the hydrolyzates hydroxyethyl-5-FU and hydroxyethyltheophylline is lower than that of 5-FU and theophylline molecules themselves. Polymeric drugs prepared by the polymerization of methacryloyl or acryloyl type monomers having 5-FU or theophylline units release them as shown in Scheme 5, therefore they are expected to be quite useful for chemotherapy. In Fig. 8 is shown the degree of hydrolysis of poly(MAFU), poly(AFU-*co*-AA), poly(AFU-*co*-AAm) and poly(MAFU-*co*-AAm) at 37 °C in a 0.1 M phosphate buffer solution as a function of time. The hydrolyzates, 5-FU, was analyzed by HPLC system with ODS-silica gel after being filtered through a Chromato Disk. As shown in this figure, 5-FU is released from backbone polymers under mild conditions. In the case of poly(MAFU), though the polymer

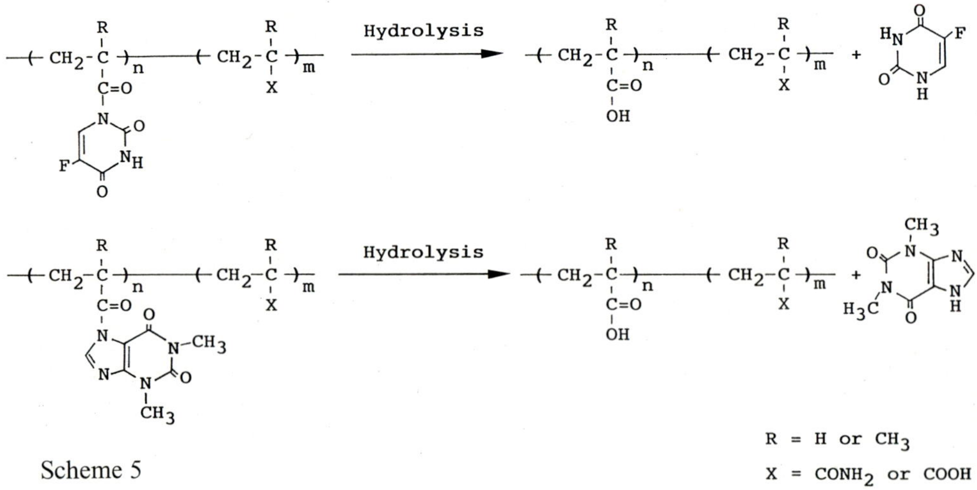

Scheme 5

R = H or CH$_3$
X = CONH$_2$ or COOH

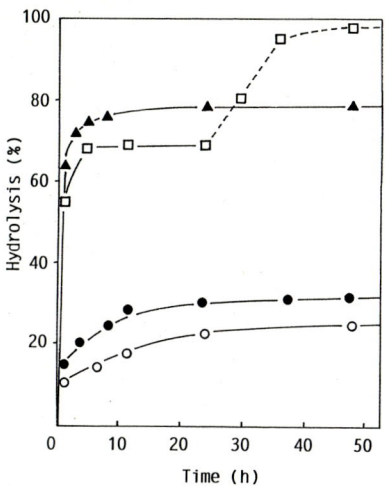

Fig. 8. Hydrolysis of polymeric drugs containing 5-FU at 37 °C in 0.1 M phosphate buffer: (□) poly(AFU-co-AAm), (▲) poly(AFU-co-AA), (●) poly(MAFU-co-AAm), (○) poly(MAFU); (———) pH 7.4, (– – –) pH 12.5

was not dissolved but dispersed in a solution, the hydrolysis in a heterogeneous system was studied. The water soluble copolymers were found to be hydrolyzed faster than the insoluble homopolymer which is similar to the methacryloyloxyethyl type polymeric drugs. The degree of hydrolysis depended on the chemical structure of drug carriers and hydrolytic conditions. The order of the degree of hydrolysis for the different chemical structures of the polymers, poly-(AFU-*co*-AAm) > poly(AFU-*co*-AA) > poly(MAFU-*co*-AAm) > poly(MAFU), may be ascribed to the hydrophilicity and flexibility of the polymers in aqueous solution.

In order to evaluate the hydrolytic behavior of polymeric drugs having theophylline, the degree of hydrolysis of AThe, MAThe, poly(AThe) and poly-

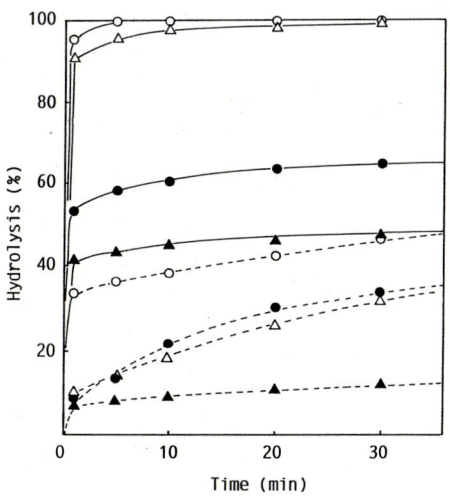

Fig. 9. Hydrolysis of polymeric drugs containing theophylline at 28 °C in 0.1 M phosphate buffer-dioxane mixture (1/3, v/v). (○) AThe, (△) MAThe, (●) poly(AThe), (▲) poly(MAThe); (———) pH 7.0, (– – –) pH 9.0

(MAThe) at 28 °C in a 0.1 M phosphate buffer-dioxane mixture (1/4, v/v) as a function of time was studied. The concentration of theophylline was determined by optical density reading with a UV spectrometer at 273 nm. The results are shown in Fig. 9. In a neutral solution, the polymeric prodrugs released theophylline molecules relatively slowly as compared to the rate at pH 9. Although AThe and MAThe were gradually hydrolyzed by atmospheric moisture, their polymeric forms were found to be more stable and released theophylline more slowly in aqueous media. The hydrolysis of methacryloyl type monomer and polymer are slower than acryloyl type derivatives, which suggests that the chemical structure of the polymeric drugs containing theophylline is important for the controlled release of pharmaceutically active compounds.

The polymeric drugs prepared from VBFU can also be expected to release 5-FU by amide hydrolysis as shown in Scheme 6. The hydrolysis of poly(VBFU) and poly(VBFU-co-MAn) was observed in a phosphate buffer solution (pH 7.0)

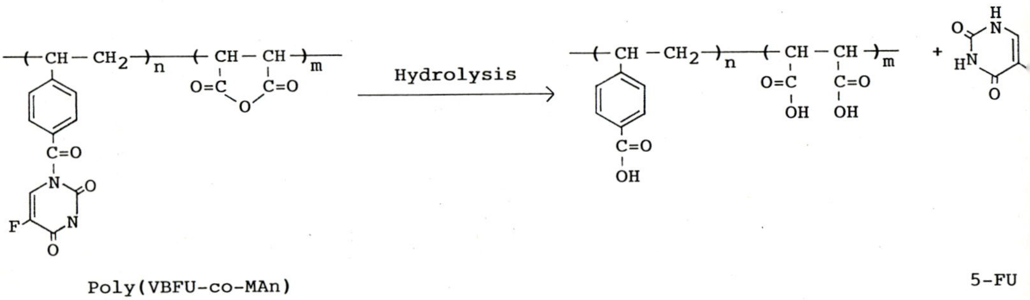

Scheme 6

and a physiological saline solution at 37 °C (Fig. 10). Since both polymers were not soluble in water, they were dispersed in water and the amount of 5-FU released was determined by HPLC with a UV (260 nm) detector after removing any undissolved polymer by filtration. Both polymeric drugs released 5-FU gradually under mild conditions (Fig. 10). In a physiological saline solution, these

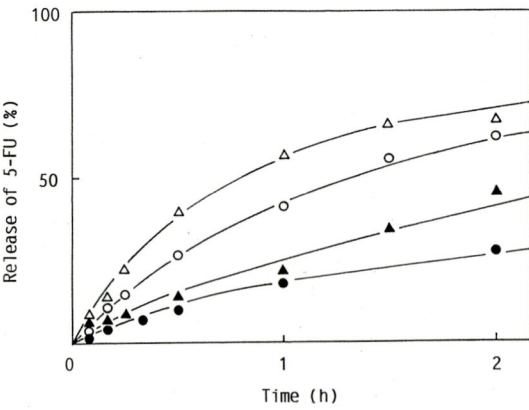

Fig. 10. Hydrolysis of polymeric drugs having 5-FU at 37 °C: (○) poly(VBFU), (△) poly(VBFU-co-MAn) in a 0.1 M phosphate buffer solution (pH 7.0), and (●) poly(VBFU), (▲) poly(VBFU-co-MAn) in a physiological saline solution

polymeric prodrugs released 5-FU more slowly than in a phosphate buffer solution. The copolymer with maleic anhydride had a tendency to release 5-FU more rapidly than poly(VBFU), but slower than poly(MAFU-co-MAn). The hydrophilicity of polymers and copolymers in aqueous solution could be an important factor in ester or amide hydrolysis of polymer side chains [4–6, 22]. should be noted that the copolymer formed after hydrolysis of poly(VBFU-co-MAn) has a typical polyanionic structure. It has been demonstrated that synthetic polyanionic materials such as the sodium salt of poly(vinyl sulfonate) and divinyl ether-maleic anhydride copolymer (pyran) possess biological activity, such as antitumor activity, through immune stimulation such as interferon-inducing activity [38]. Therefore, poly(VBFU-co-MAn) could be expected to show biological activity even after complete release of 5-FU. Actually, it has been observed that poly(styrene-co-MAn) (SMA) exhibits significant antitumor activity [41]. Recently, Maeda et al. found that a SMA conjugated to the anticancer protein neocarzinostatin (SMANCS) accumulated more in tumor tissues than did neocarzinostatin [42]. The present poly(VBFU-co-MAn) has a very similar structure to SMA. The polyanion itself, may be effective against the tumor by an immune mechanism similar to divinyl ether-maleic anhydride copolymer (pyran) and the other polyanion [43].

2.5 Antitumor Activity of Polymeric Drugs Having 5-Fluorouracil

The antitumor activity of synthetic 5-FU derivatives, polymeric prodrugs having 5-FU unit, and physically 5-FU incorporated polymers has been investigated [44–55]. Moreover recent attention has been focused on syntetic polyanionic polymers [41], and drug-polyanion conjugates have been prepared to evaluate the activities [56].

The antitumor activity of poly(MAFU-co-MAn), poly(VBFU), and poly-(VBFU-co-MAn) was examined using tumorbearing mice. Four-week-old male albino mice of dry strain weighing about 25 g were injected with Ehrlich's ascites tumor cells or subcutaneously implanted with Sarcoma 180 ascites tumor cells. Poly(MAFU-co-MAn), poly(VBFU), and poly(VBFU-co-MAn) were dispersed in water and daily administrated by intraperitoneal injection along with 5-FU (0.4 mg) for 7 days. The total packed cell volume (TPCV) was measured on Ehrlich's ascites tumor bearing-mice after treatment. The mice were sacrificed after 8 days, and the ascites cells were collected into a hematocrit tube. The

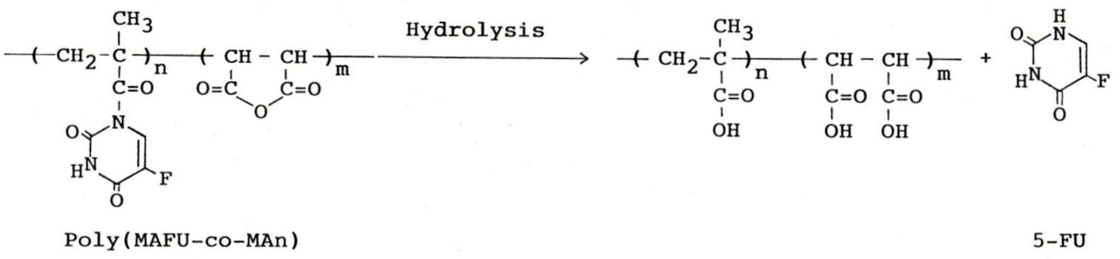

Poly(MAFU-co-MAn)   5-FU

Scheme 7

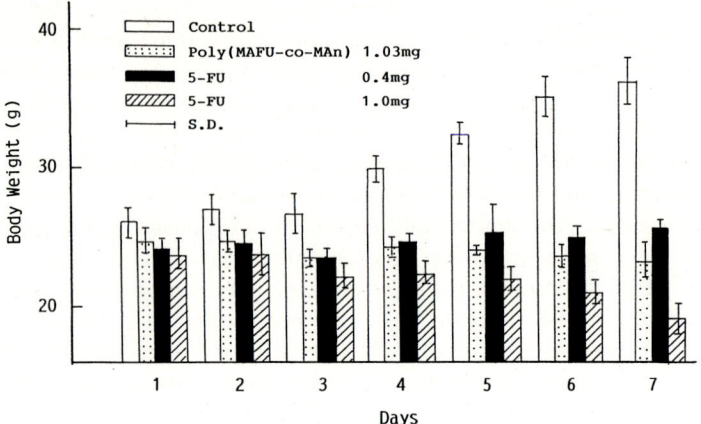

Fig. 11. Changes in the body weight of Ehrlich's ascites tumor-bearing mice after treatment with antitumor drugs

tube was centrifuged (11,000 rpm, 5 min) in order to determine the volume ratio of tumor cells in ascites (PCV). The ascites volume as change of the body weight before and after removal the ascites multiplied by PCV was TPCV. Survival time was observed among number of the dead after daily administration by intraperitoneal injection for 7 days using 4 mice in a group. For the purpose of estimating of antitumor activity against solid type tumor, Sarcoma 180 ascites tumor cells were implanted subcutaneously in 8 mice. After polymeric drugs were administered by intraperitoneal injection for 7 days, the mice sacrificed after 11 days, the subcutaneous tumor was excised from the mice and the tumor weight and surface area were measured.

The results based on changing body weight after injection of poly(MAFU-co-MAn), which contains 58 mole-% of MAFU, into the tumor-bearing mice are

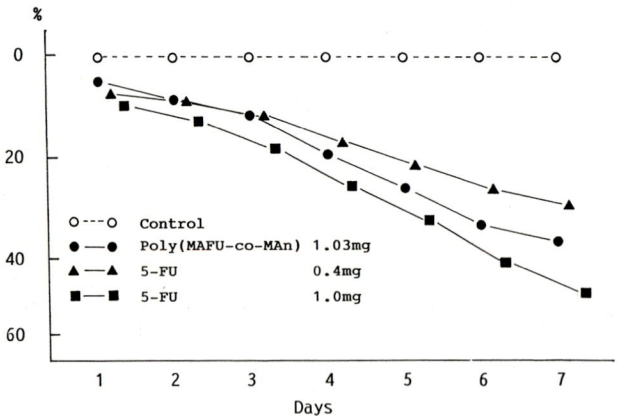

Fig. 12. Change in the body weight of Ehrlich's ascites tumor-bearing mice by treatment with antitumor drugs compared to untreated animals

shown in Figs. 11 and 12. Poly(MAFU-co-MAn) (1.03 mg) was dispersed in water and daily administrated by intraperitoneal injection along with 5-FU (0.4 mg or 1.0 mg) for 7 days. As shown in Fig. 11, both poly(MAFU-co-MAn) and 5-FU were effective against Ehrlich's ascites. The tumor-bearing mice which were treated showed no change in body weight. In order to clarify the effect of the copolymer, the percentage difference in body weight of mice after treatment was compared to untreated mice (Fig. 12). The copolymer appears to be more effective against tumor-bearing mice than an equivalent amount of free 5-FU (0.4 mg), but less effective than that of 1.0 mg of free 5-FU.

To determine whether the chemicals acted on the tumor cells, the total packed cell volume (TPCV) was measured on tumor-bearing mice after treatment with poly(MAFU-co-MAn) and 5-FU itself. Shown in Table 9 are the results of TPCV after 7 days. The presence of ascites in these mice was not apparent after using either poly(MAFU-co-MAn) or 5-FU, which suggests that the present polymeric drug may directly affect the ascites tumor cells in a similar manner to 5-FU. In conclusion, poly(MAFU-co-MAn) was found to be effective against Ehrlich's Ascites tumor cells like 5-FU. The antitumor activity is greater than an equivalent amount of 5-FU alone. This may be attributed to the covalently bound 5-FU. The reason for the activity of the present polymeric drug may be as follows: 1) Sustained release of 5-FU from the polymer chain is effective. 2) The polyanion formed after hydrolysis may be effective against the tumor by an immune mechanism similar to pyran and the other polyanion [41]. 3) The initial copolymers are directly incorporated into tumor cells or macrophages by

Table 9. Total packed cell volume (TPCV) of Ehrlich's ascites tumor-bearing mice after 7 days

Exp. No.	Antitumor Drugs	Ascites (ml)	mean PCV* (%)	TPCV**
1	1.03 mg of poly(MAFU-MAn)	0	0	0
2		0	0	0
3		trace	0	0
4	0.4 mg of 5-FU	0	0	0
5		0	0	0
6		trace	0	0
7	1.0 mg of 5-FU	trace	0	0
8		trace	0	0
9		0	0	0
10		trace	0	0
11	Control	5.1	36	1.836
12		8.4	24	2.016
13		5.7	43	2.423

* PCV = packed cell volume
** TPCV = total packed cell volume

Table 10. Antitumor activity of 5-FU and vinyl-type polymeric drugs against Ehrlich's ascites tumor bearing mice[a]

Drug (mg)	% Inhibition of body weight	TPCV[b] (mean + S.E.)	Survival Time[c] (days, mean + S.E.)	ILS[c]
Control (none)	0	2.1 ± 0.1	16.4 ± 1.2	0
5-FU (0.40)	13	0	33.0 ± 3.9	101
Poly(VBFU) (0.81)	15	0.4 ± 0.2	25.5 ± 4.3	55
Poly(VBFU-co-MAn) (0.96)	16	0.3 ± 0.2	23.8 ± 1.3	45

[a] Inoculated cells; $9.0 \times 10^6/0.1$ mL i.p.
[b] Total packed cell volume.
[c] Inoculated cells; $6.6 \times 10^6/0.1$ mL i.p.

phagocytosis, since they are administered as a fine powder resembling microspheres [40].

The antitumor activity of poly(VBFU) and poly(VBFU-co-MAn) was evaluated by the change of the body weight, total packed cell volume (TPCV), and survival time with Ehrlich's ascites tumor-bearing mice, and the tumor weight and surface area for subcutaneous implants of Sarcoma 180 ascites tumor-bearing mice. Since both polymeric drugs were insoluble in water, they were dispersed in water and then administrated by intraperitoneal injection along with 5-FU. The results of the antitumor activity against Ehrlich's ascites tumor-bearing mice were summarized in Table 10. Both polymeric drugs and the equimolar amount of 5-FU inhibited increase of body weight after treatment for 7 days. The chemicals showed a significant depression in TPCV, indicating that the released 5-FU and/or hydrolyzed polyanion affected the ascites tumor cells. Similar inhibition effect of body weight was also observed in the case of poly(MAFU-co-MAn) upon the same tumor-bearing mice. The polymeric drugs induced a high increase in life span percentage (ILS) (ca. 50%) although the equimolar 5-FU showed a more efficient prolongation effect. Shown in Table 11 are the results of antitumor activity against the subcutaneous type of Sarcoma 180 ascites tumor-bearing mice. Only poly(VBFU-co-MAn) showed an activity upon the subcutaneous tumor.

Table 11. Antitumor activity of 5-FU and polymeric drugs against subcutaneous type of Sarcoma 180 ascites tumor cells[a].

Drug (mg)	Tumor weight (g)[b]	Surface area (mm²)[b]
Control (none)	0.98 ± 0.24	148 ± 48
5-FU (0.40)	1.50 ± 0.29	201 ± 42
Poly(VBFU) (0.81)	2.36 ± 0.54	293 ± 65
Poly(VBFU-co-MAn) (1.07)	0.70 ± 0.17	45 ± 13

[a] Inoculated cells; $1.12 \times 10^7/0.1$ mL s.c.
[b] Mean + S.E.

The chemical structure of polymeric drugs seems to be very important for the antitumor activity, while the relationship between 5-FU release and antitumor activities has not been clarified. Antitumor activity of the other polymeric drugs having 5-FU is now in progress.

3 Nucleic Acid Analogs and Polyanionic Polymers as Polymer Drugs

3.1 Polymer Synthesis

Among the nucleic acid analogs prepared so far, vinyl type polymers bearing nucleic acid bases such as poly(VAd) are known to form a strong complex with a naturally occurring nucleic acid in an aqueous medium [57–59]. As shown in Scheme 8, 9-vinyladenine (VAd) monomer was prepared conveniently in a high

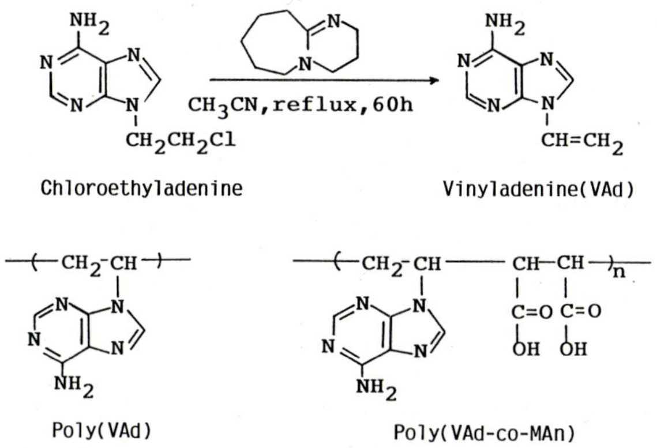

Scheme 8   Poly(VAd)   Poly(VAd-co-MAn)

yield from chloroethyladenine by dehydrochlorination. Free radical polymerization of VAd was carried out in dimethylsulfoxide at 95 °C, while the copolymerization of VAd and MAn was done at 60 °C in acetonitrile or dioxane. The results are summarized in Table 12. Both polymers obtained were water soluble enough for their biological activities to be evaluated.

Table 12. Polymerization of vinyladenine (VAd) at 95 °C and copolymerization of VAd and maleic anhydride at 60 °C

Run No.	Monomer (mol%)		Solvent	Conversion (wt%)	m_1 in copolymer
	M_1	M_2			
1	VAd (100)		DMF	17	—
2	VAd (100)		DMSO	31	—
3	VAd (33)	MAn (67)	acetonitrile	12	61
4	VAd (33)	MAn (67)	dioxane	28	58

Total concentrations of the monomer were 2.0×10^{-1} M in runs 1 and 2, while those were 3.0×10^{-1} M in runs 3 and 4. [BPO] = 1.0×10^{-3} mol/L

3.2 Interactions

In order to clarify the solution properties of poly(VAd) in more detail, poly(VAd) was fractionated by ultrafiltration techniques. The observed molar extinction coefficient of poly(VAd) apparently decreases with an increase in the average molecular weight (Table 13). This suggests that the possibility of an inter-

Table 13. Molar extinction coefficient as observed by changing the molecular weight of poly(VAd)

M_w	Absorbance (cm^{-1} M^{-1})	Hypochromicity$_I$ (%)[a]
$M_w < 500$	8400	24
$500 < M_w < 1000$	7100	36
$1000 < M_w < 5000$	5900	47
$5000 < M_w < 10{,}000$	5300	52
$10{,}000 < M_w$	4900	56

[a] Calculated on the basis of the value of 9-ethyladenine.

molecular or intramolecular interaction of the adenine moieties in the polymer. This can be explained in terms of "Hypochromism" [60]. The "Hypocromicity$_I$" is calculated by Eq. (1) (Table 13).

$$\text{Hypochromicity}_I(\%) = [1 - \varepsilon_{\text{polymer}}/\varepsilon_{\text{monomer unit}}] \times 100 \tag{1}$$

In a physicochemical study of oligo- or polynucleotides, it has been reported that "Hypochromicity$_I$" depends on the degree of polymerization [61]. The value of native DNA is about 60%, which corresponds to the "Hypochromicity$_I$" of the absorption of its mononucleotides [60]. The present result indicates that the purine rings may intramolecularly stack by themselves in an aqueous solution

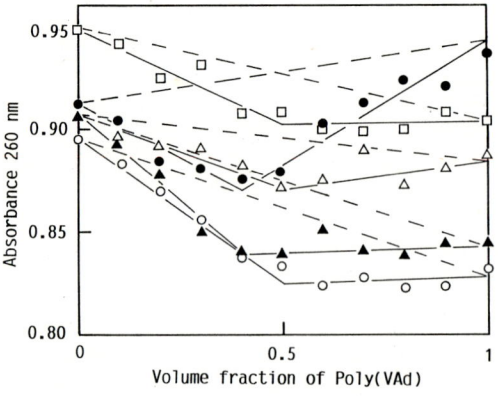

Fig. 13. Mixing curve between poly(VAd) and RNA in H$_2$O (pH 7.0). Absorbance was determined at 260 nm. A mixed solution was allowed to stand at room temperature for 3 h

Table 14. Hypochromicity$_{II}$ (%) of the complex formed between poly(VAd) and RNA as a function of the molecular weight

Poly(VAd) M_w	Hypochromicity$_{II}$ (%)
$M_w < 500$	2.6
$500 < M_w < 1000$	3.0
$1000 < M_w < 5000$	4.4
$5000 < M_w < 10,000$	4.6
$10,000 < M_w$	6.1

[30, 62] in a manner similar to polynucleotides [61] and that the higher molecular weight poly(VAd) has a more stable conformation.

Figure 13 shows the mixing curve [4, 9–11, 57, 63, 64] for the intensity change as a function of the mixing ratio between synthesized nucleic acid analogs and naturally occurring RNA in an aqueous solution as observed by Imoto et al. [59]. The "Hypochromicity$_{II}$" when two different polymers were mixed was calculated using Eq. (2). Results are listed in Table 14.

$$\text{Hypochromicity}_{II}(\%) = [1 - I^{a+b}/(m \times I^a + n \times I^b)] \times 100 \quad (2)$$

In Eq. (2), subscripts a and b are the interactants, and m and n are their volume fractions. I^a, I^b and I^{a+b} are the absorbance for solution of a, b and a + b mixture, respectively. The "Hypochromicity$_{II}$" increases with an increase in the molecular weight of poly(VAd). The "Hypochromism" in polynucleotides and nucleic acid analogs is generally caused by stacking of nucleotides in the polymer side chains [60]. In the case of nucleic acid analogs, several types of complex formation may be considered: (1) the complex formation by polymer-polymer intermolecular interaction [Scheme 9 (a) and (b)], (2) the intramolecular complex formation in a folded polymer [Scheme 8 (c)], and (3) the complex formation of complementary base polymers [Scheme 9 (d)]. As shown in Scheme 9 (a) and (b), the intermolecular hydrogen bonding of nucleobase or base stacking may be the driving force for complex formation. If the polymers

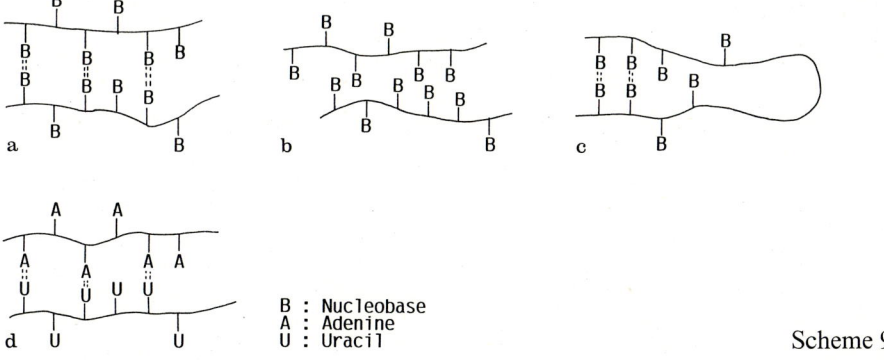

Scheme 9

are flexible enough, the intramolecular interaction can take place as in the case of tRNA. When polymers having uracil moieties, such as poly vinyluracil [58], polyU [57] or RNA [59], are mixed with poly(VAd), the complex formation as in Scheme 9 (d) is expected. For the polyA-polyU complex [65] the chemical structure and hypochromism of the complex are well-known. However, the poly(VAd)-RNA complex has not been evaluated. The present "Hypochromicity$_{II}$" results suggest that poly(VAd) with high molecular weight interacts with RNA relatively strongly. For the high molecular weight poly(VAd) it is assumed that the amine groups of the adenine moieties face the aqueous phase. Therefore, poly(VAd) can form the complex with RNA by hydrogen bonding. Thus intramolecular base-base stacking of poly(VAd) and/or RNA could be enhanced causing an absorption decrease. However, the alternating copolymer of VAd and maleic acid (abbreviated as poly(VAd-co-MAn)), which was obtained by the copolymerization of VAd with MAn and hydrolysis, did not show any change in its absorption spectra upon the addition of RNA. This suggests that anionic moiety of poly(VAd-co-MAn) may play an important role in the complex formation.

Complex formation between RNA and water soluble copolymers obtained in the copolymerization of methacryloyloxyethyl-type monomers containing nucleobases with water soluble monomers was also studied. Mixing curves between copolymers and between copolymers and RNA are shown in Figs. 14–16. The interaction between poly(MAOFU-co-AAm), poly(MAOT-co-AAm), or poly-(MAOA-co-AAm) with RNA was observed, as shown in Fig. 14. The overall stoichiometry of the complexes was about 1:1 and the hypochromicity was about 2% for the copolymer-RNA system under the conditions used. The observed interaction was not as strong as for the poly(VAd)-RNA system and poly(MAOA)-poly(MAOT) system [64], since the solubilizer, AAm, in the

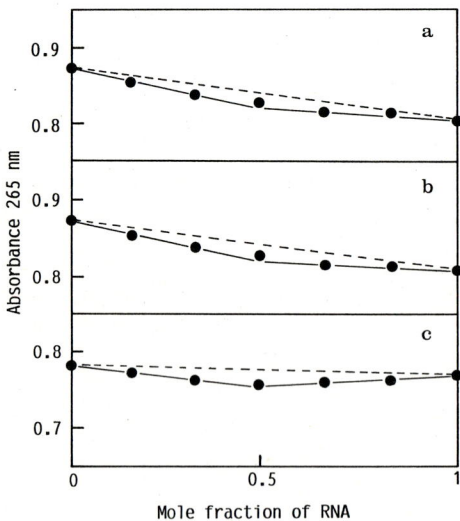

Fig. 14a–c. Mixing curve between copolymers and RNA in water. Absorbance at 265 nm obtained in a 10-mm cell at 20 °C. (a) Poly(MAOFU-co-AAm)-RNA system, (b) poly(MAOT-co-AAm)-RNA system, (c) poly(MAOA-co-AAm)-RNA system

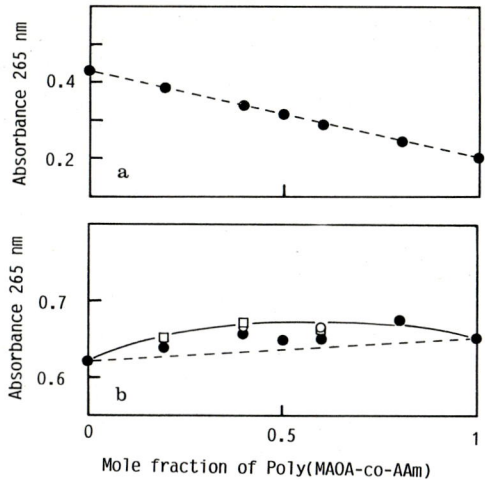

Fig. 15a, b. Mixing curve between poly-(MAOT-co-AAm) and poly-(MAOA-co-AAm). Absorbance at 265 nm obtained in a 10-mm cell at 20 °C. (a) In water, (b) in DMSO-ethylene glycol mixture (3/2, v/v): (●) after 5 h, (○) after 15 h, (□) after 30 h

copolymers interrupts the base-base stacking in the polymers side chain and weakens complex formation.

Figure 15 shows the interaction between poly(MAOT-co-AAm) and poly-(MAOA-co-AAm) in water and DMSO-ethylene glycol, respectively. No hypochromicity was observed in water, while hyperchromicity [67] was observed in DMSO-ethylene glycol. These results can be explained by assuming that the nucleic acid bases are isolated along the copolymer chain and difficult to stack intramolecularly, hence it is easy to form complementary base pairs between copolymers through hydrogen bonding. Generally, hydrophobic interactions by base-base stacking are predominant in aqueous solution, and hydrogen bonding interactions have been observed in organic solvents [64]. This fact may support the present experimental results.

Figure 16 shows the mixing curve between copolymers containing AA and RNA in water. In aqueous solution, purine and pyrimidine bases in co-

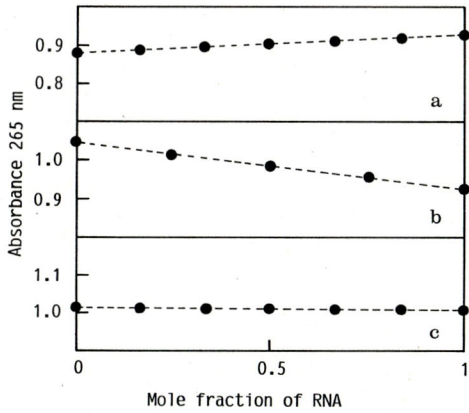

Fig. 16a–c. Mixing curve between copolymers and RNA in a 0.1 M phosphate buffer (pH 7.8). Absorbance at 265 nm obtained in a 10-mm cell at 20 °C. (a) Poly-(MAOFU-co-AA)-RNA system, (b) poly-(MAOT-co-AA)-RNA system, (c) poly-(MAOA-co-AA)-RNA system

polymers can interact by base-base stacking as shown in Fig. 14. However, even for copolymers with high contents of nucleic acid bases such as poly-(MAFU-co-AA), poly(MAOT-co-AA) and poly(MAOA-co-AA), no interaction was observed. This is explained by the repulsion between electronegative charges of carboxyl groups in the copolymers and those of phosphate group in RNA, thus preventing a mutual approach of nucleic acid bases.

SMA, which is obtained by the free radical alternative copolymerization of styrene and maleic anhydride followed by hydrolysis and partial alkylation, is a water soluble synthetic polyanionic polymer having hydrophobic moieties and has biological activities [67]. Figure 17 shows the mixing curves between yeast RNA and SMA (Fujiyoshi & Co., Ltd; SMA 1000) in comparison with that between RNA and poly(VAd) (Mn > 50,000). As shown in Fig. 17 (a) and (b), hypochromicity by mixing at pH 7.0 and after standing for a couple of hours, was observed at a relatively high concentration, though the interaction is weaker than that seen in the complex between poly(VAd) and RNA (Fig. 17 (c)).

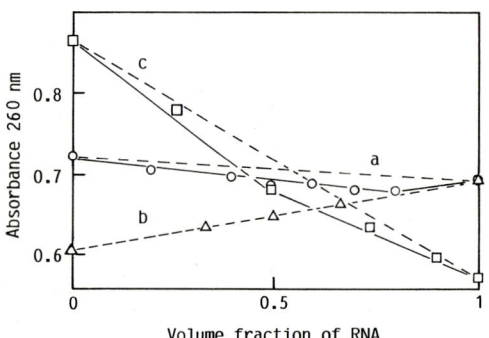

Fig. 17. Mixing curve between RNA and SMA or poly(VAd) in a 0.1 M phosphate buffer solution at pH 7.0. Absorbance at 260 nm was obtained at room temperature. (a) — ○ —, RNA-SMA system in a 2-mm cell, (b) — △ —, RNA-SMA system in a 10-mm cell, (c) — □ —, RNA-poly(VAd) system in a 2-mm cell

Generally, the complex between nucleic acid analogs, syntetic or natural polynucleotides is caused by the hydrogen bonding interaction between complementary nucleobases in aqueous solution. Since SMA has no functional group for hydrogen bonds, the interaction observed can be explained by the hydrophobic interaction between nucleobase of RNA and benzene rings of SMA. It is throught that SMA plays a role as a polymer intercalating agent.

Consequently, both nucleic acid analogs such as poly(VAd) and pharmaceutically active syntetic polyanionic polymers having hydrophobic moieties may interact with nucleic acids in vivo.

3.3 Encapsulation of Poly(vinyladenine) and Poly(vinyladenine-alt, co-maleic acid) in Liposome

Most of the nucleic acid model compound prepared so far are water insoluble; however, when the analogs are water soluble, they may not permeate into the hydrophobic cell membranes. Recently, an improved drug delivery system for water soluble drugs using polysaccharide-coated liposomes [68] was developed.

Thus, if one can provide a water soluble polymer drug, it can be encapsulated into specific liposomes for pharmaceutical delivery.

To encapsulate water soluble nucleic acid analogs, i.e. poly(VAd) and poly(VAd-co-MAn), in modified liposomes to effectively transfer the polymers into the target cells [69], we must consider the lysis of liposomal membrane during the procedures of encapsulation. In order to examine whether poly(VAd) and poly(VAd-co-MAn) can be stably encapsuled in the liposomes, the interaction between the liposomal membranes and the polymers was first investigated. A thin film of egg phosphatidylcholine was dispersed in an aqueous 20 mM Tris-HCl buffer solution (pH 8.6) containing 200 mM NaCl and 200 mM carboxyfluorescein (CF). The resulting suspension was sonicated under nitrogen atmosphere to give conventional small unilamellar liposomes (SUV) with CF in the interior water phase [69]. The CF-loaded SUV were separated from the uncapsuled CF by gel-filtration on a Sepharose 4B column with Tris-HCl buffer as the eluent. An aqueous solution of poly(VAd) was added at a given concentration to the liposomal suspension after an appropriate preincubation at 37 °C. Any perturbation to the liposomal membrane would cause CF-release from the liposomes and emit a strong fluorescence at 520 nm. Results of the CF-release induced by the addition of poly(VAd) as a function of time at 37 °C and 50 °C are shown in Fig. 18. Clearly, at both temperatures the CF-release was not significant. Therefore poly(VAd) can be effectively encapsulated into liposomes.

Shown in Fig. 19 are the results of CF-release induced by poly(VAd-co-MAn) at 37 °C and 50 °C, respectively. At 50 °C, CF was released by the addition of the copolymer, particularly for the high molecular weight polymers. At 37 °C, it is less serious compared with previous results observed for polyanionic polymers [69].

The mechanism of macrophage activation with polymer drugs such as synthetic polynucleotides and polyanionic polymers is not clearly understood. Two possibilities based on physiological responses are: (1) the direct interaction or perturbation by the polymer drug with the cell cytoplasma membrane, and (2)

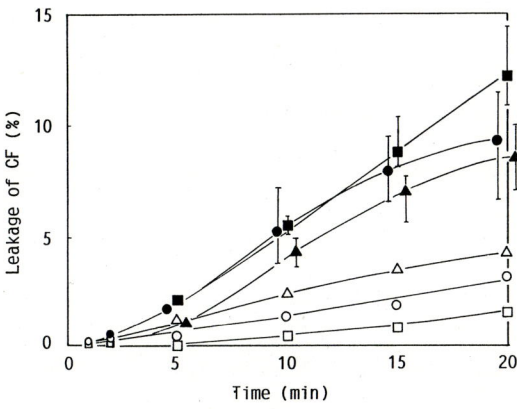

Fig. 18. CF-Release from liposome upon the addition of poly(VAd) (0.5 mg/mL) at 37 °C and 50 °C. [Egg PC] = 1.3×10^{-4} M, — ○ — control at 37 °C, — □ — poly(VAd) (10,000 < MW < 50,000) at 37 °C, — △ — poly(VAd) (50,000 < MW < 100,000) at 37 °C, — ● — control at 50 °C, — ■ — poly(VAd) (10,000 < MW < 50,000) at 50 °C, — ▲ — poly-(VAd) (50,000 < MW < 100,000) at 50 °C

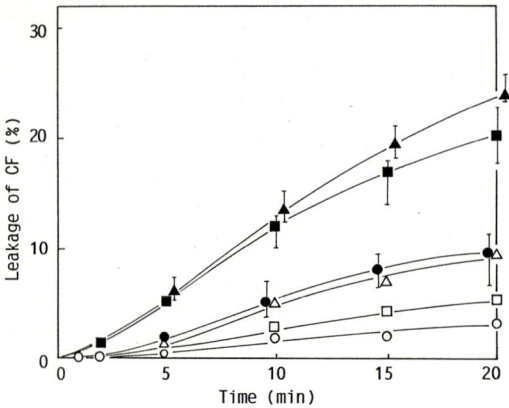

Fig. 19. CF-Release from liposome upon the addition of poly(VAd-co-MAn) (0.5 mg/mL) at 37 °C and 50 °C. [Egg PC] = 1.3×10^{-4} M, — ○ — control at 37 °C, — □ — poly(VAd) (10,000 < MW < 50,000) at 37 °C, — △ — poly(VAd) (50,000 < MW < 100,000) at 37 °C, — ● — control at 50 °C, — ■ — poly(VAd) (10,000 < MW < 50,000) at 50 °C, — ▲ — poly(VAd) (50,000 < MW < 100,000) at 50 °C

the physiological reaction of the polymer drug with an enzyme or DNA in the cell. Our studies of the interaction between these polymer drugs and liposome as a cell model suggest that the perturbation of the cell membrane is not a significant factor compared to the activity after internalization into the cells.

3.4 Potent Immunomodulating Activity

Several pharmaceutical activities of nucleic acid analogs such as poly(VAd) have been studied in vitro and in cell-free systems [19]. It was expected that the present polymers would be effectively transferred into phagocytes by encapsulating in polysaccharide-coated liposomes and would show increased pharmaceutical activities similar to poly(maleic acid-*alt*-2-cyclohexyl-1,3-dioxap-5-ene) (MA-CDA) [68]. The activation of human neutrophils by poly(VAd) was evaluated by monitoring the in vitro superoxide anion production from activated human neutrophils. Shown in Table 15 is the superoxide liberated from human neutrophils (1×10^6 cells/ml) activated by poly(VAd) (0.5 mg/ml) as a function of time. Poly(VAd) encapsulated in mannan derivative-coated liposomes showed a

Table 15. Superoxide anion production from human neutrophils activated by poly(VAd)

	O_2-Production nmol/min/1×10^6 cells (% Increase in O^{2-} production)		
	1 h	2 h	3 h
Poly(VAd) (50 µg/100 µl) in CHM-LUV[a]	10.31 (124%)	9.95 (118%)	9.16 (117%)
Free poly(VAd) (50 µg/100 µl)	8.38 (101%)	8.59 (102%)	8.22 (105%)
Control[b]	8.32 (100%)	8.42 (100%)	7.85 (100%)

[a] Egg lecithin LUV [(egg lecithin) = 0.75 mg/100 µL] coated by CHM-200-1.8 (75 µg/100 µL).
[b] Phosphate buffered saline

Table 16. Superoxide anion production from human neutrophils activated by fractionated poly(VAd) (5000 < MW < 10,000)

	O^{2-} Production nmol/min/L × 10⁶ cells (% Increase in O^{2-} production)		
	1 h	2 h	3 h
Poly(VAd) (100 µg/100 µl) in CHM-LUV[a]	8.59 (96%)	9.70 (117%)	6.35 (113%)
Free poly(VAd) (100 µg/100 µl)	9.09 (102%)	9.55 (115%)	4.89 (87%)
Empty (PBS) in CHM-LUV	8.03 (90%)	8.97 (108%)	6.22 (111%)
Control[b]	8.93 (100%)	8.31 (100%)	5.62 (100%)

[a] Egg lecithin LUV [(egg lecithin) = 0.75 mg/100 µL] coated by CHM-200-1.8 (75 µg/100 µL).
[b] Phosphate buffered saline

greater superoxide anion production (124%) than free poly(VAd) (5,000 < MW < 10,000). The superoxide anion production reached the maximum after 2 h with no significant differences after incubation (Table 16). Superoxide anion production from human neutrophils (1 × 10⁶ cells/ml) activated by fractionated poly(VAd-*co*-MAn) (1,000 < MW < 10,000) (1.0 mg/ml) as a function of time also showed no significant difference (Table 17).

Poly(VAd) in the polysaccharide-coated liposomes activated mouse peritoneal macrophages more effectively than free poly(VAd) and the % increase in O^{2-} maximizes at 2 h after administration (Fig. 20). These results show that poly(VAd), which is a typical synthetic analog of nucleic acids [19], modulates immune responses similar to native polynucleotides [70]. Poly(VAd-*co*-MAn) in polysaccharide-coated liposome activated mouse peritoneal macrophages more effectively than free poly(VAd-*co*-MAn). The percent increase in O^{2-} production at

Table 17. Superoxide anion production from human neutrophils activated by poly(VAd-co-MAn) (1000 < M_w < 10,000)

	O^{2-} Production nmol/min/L × 10⁶ cells (% Increase in O^{2-} production)		
	1 h	2 h	4 h
Poly(VAd-co-MAn) (100 µg/100 µL) in CHM-LUV[a]	12.96 (127%)	8.85 (95%)	7.67 (75%)
Free poly(VA-co-MAn) (100 µg/100 µL)	12.96 (127%)	7.08 (76%)	6.67 (66%)
Control[b]	10.24 (100%)	9.28 (100%)	10.18 (100%)

[a] Egg lecithin LUV [(egg lecithin) = 0.75 mg/100 µL] coated by CHM-200-1.8 (75 µg/100 µL).
[b] Phosphate buffered saline

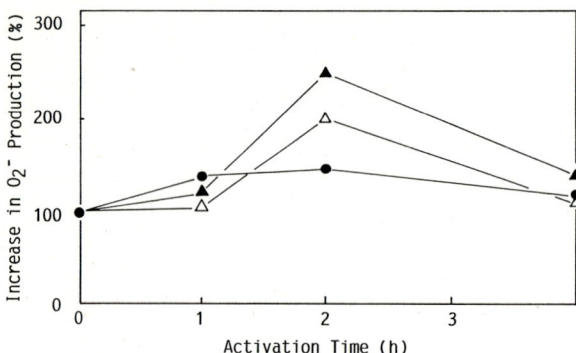

Fig. 20. Superoxide anion production from murine peritoneal macrophage after in vivo activation with poly(VAd) encapsulated in the mannan derivation-coated LUV. — ▲ — Poly(VAd)/LUV, — ● — free poly(VAd), — △ — empty LUV without poly(VAd)

2 h after i.p. injection is shown in Table 18. A drastic increase (490%) was observed when poly(VAd-*co*-MAn) was encapsuled in the polysaccharide-coated liposome. This is similar to other several polyanionic polymers, such as MA-CDA [68] or SMANCS [67] which have been found to exhibit biological activities in vivo.

The present investigation shows that the combination of polyanion moieties and a nucleobase, such as poly(VAd-*co*-MAn) or poly(VAd), can increase immunodulating activity, particularly in vivo. Although the interaction of synthetic polymer immunomodulators and nucleic acids in the nucleus of cells might be one of the important factors for the manifestation of activity [19], we could not find any direct correlation between the interaction of synthetic polymers with RNA in vitro and biological activities in vivo. Further studies on the molecular weight dependency of the polymer on the biological activity and the effect of the chemical structure on the immunomodulating activity are in progress.

Table 18. Superoxide anion production from mouse macrophage activated by poly(VAd-co-MAn) 2 h after i.p.

	O^{2-} Production nmol/min/2×10^5 cells	% Increase in O^{2-} Production
Poly(VAd-co-MAn) (150 μg/100 μL) in CHM-LUV[a]	1.575 ± 0.055	490
Poly(VAd-co-MAn) (100 μg/100 μL) in CHM-LUV	1.250 ± 0.040	380
CHM-LUV	0.801 ± 0.055	249
Poly(VAd-co-MAn) (150 μg/100 μL)	1.285 ± 0.135	377
Poly(VAd-co-MAn) (100 μg/100 μL)	0.779 ± 0.006	248
Control[b]	0.317 ± 0.018	100

[a] Egg lecithin LUV [(egg lecithin) = 0.75 mg/100 μL] coated by CHM-200-1.8 (75 μg/100 μL).
[b] Phosphate buffered saline

4 Drug-Immobilized Biomaterials

4.1 Immobilization of Heparin and/or Prostaglandin-I$_2$

Blood vessels contribute to fluent blood circulation by preventing thrombosis. A major barrier to thrombus formation in vivo is the intact endothelial lining of blood vessels; normal endothelium neither activates coagulation nor allows platelet adhesion. The nonthrombogenic activity of endothelium is accomplished in part by the production of the icosanoid mediator PGI$_2$. PGI$_2$ acts as a vasodilator and inhibits platelet activation [71]. Recently, other endothelial factors that inhibit coagulation reactions have been described. These include heparin-like molecules on the endothelial cell surface that act as a cofactor for antithrombin III in inhibiting several coagulation factor proteases [72], and the cell surface protein, thrombomodulin, that converts thrombin to an anticoagulant from procoagulant enzyme [73].

Blood-compatible polymer materials are required to inhibit both platelet adhesion and coagulation just as the endothelial on the polymer surface. It is known that there are many investigations in the design and the synthesis of socalled antithrombogenic materials. The immobilization of biologically active substances such as heparin [74, 75], urokinase [76], and prostaglandins [77–81] is one of the practical approaches.

Immobilization of biologically active molecules may be classified into the following two categories: (1) immobilization by chemical bonds, (2) immobilization by physical entrapment. It is also known that both heparinized-surface and released-heparin are effective for antithrombogenicity of biomaterials. When biomaterials are required to be antithrombogenic for a relatively short term, the slow release of antithrombogenic materials may be extremely effective. Among polymer materials poly(vinyl chloride) (PVC) has good mechanical characteristics for biomedical use, therefore it is currently one of the important commercially available biomaterials [82]. However, the blood compatibility of PVC is not satisfactory for medical use. Surface modified and heparinized PVC's [83–86] have been prepared in order to achieve antithrombogenicity and are being used for blood bags and catheters. Since heparin is not soluble in tetrahydrofuran, which is a good solvent for PVC, but very soluble in water, it is difficult to directly immobilize in the PVC matrix. However, heparin is easily entrapped in hydrogels such as crosslinked poly(acrylamide) gel by the copolymerization of AAm and Bis A in the presence of heparin in aqueous solution under mild conditions. The entrapped heparin is slowly released from the hydrogel.

Initially heparin was physically entrapped in polyacrylamide gels by a bead polymerization technique in a water/benzene mixture using detergent. The results are shown in Table 19. The hydrogels were freeze-dried and the dry gels with 0.1–1.10 μm diameter were dispersed in a THF solution containing PVC and cast into a film. Shown in Fig. 21 is a scanning electron micrograph of the surface of PVC containing hydrogels. As seen in this figure, the hydrogels were well dispersed in the surface and the interior of the PVC film.

Shown in Fig. 22 are the stress-strain curves, tensile strength and elongation

Table 19. Preparation of hydrogels with heparin or PGI$_2$

Sample No.	Monomer [g, (mmol)]					Heparin (mg)	PGI$_2$ (mg)
	AAm	DMAPAA	NMAA	DMAA	bis (A)		
1	0.094 (1.32)				0.001 (0.007)	50	none
2	0.094 (1.32)				0.001 (0.007)	150	none
3	0.047 (0.66)				0.0005 (0.003)	250	none
4	1.414 (19.9)				0.015 (0.10)	45	none
5	0.347 (4.88)	0.0016 (0.01)			0.0039 (0.025)	23	none
6			0.847 (9.95)		0.0077 (0.05)	23	none
7				0.986 (9.95)	0.0077 (0.05)	23	none
8	0.707 (9.95)				0.0077 (0.05)	none	none
9	0.707 (9.95)				0.0077 (0.05)	none	1.0
10			0.847 (9.95)		0.0077 (0.05)	none	1.0
11				0.986 (9.95)	0.0077 (0.05)	none	1.0

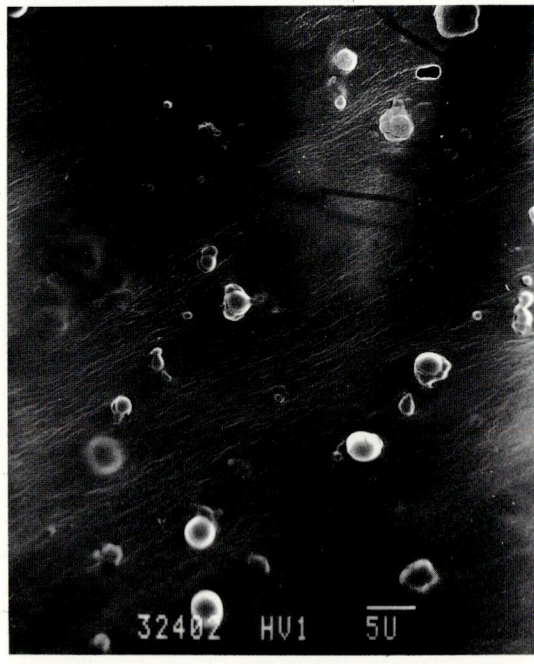

Fig. 21. Scanning electron micrograph of the surface of PVC with hydrogels

New Aspects of Polymer Drugs

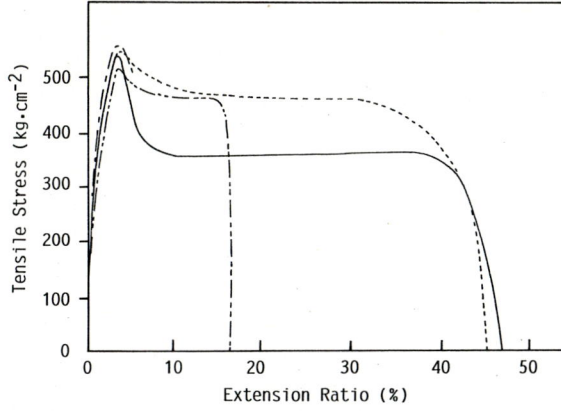

Fig. 22. Stress-strain curves of PVC with hydrogels. (—·—) PVC with hydrogels (Sample No. 4 in Table 19, (—---—) PVC with hydrogels (Sample No. 6), (— — —) PVC with hydrogels (Sample No. 7), (———) PVC

of the PVC films containing hydrogels. The tensile strength of the PVC films is not weakened very much by the addition of the hydrogels. The hydrogel consisting of *N,N*-dimethylacrylamide (DMAA) exhibits an elongation which suggests that the compatibility of alkyl groups of DMAA with PVC may be of importance for mechanical strength.

4.2 Release of Heparin

In order to estimate the release of heparin from the PVC matrix in blood, the amount and the rate of release of heparin from the PVC matrix were examined at 37 °C in physiological saline. Figure 23 shows the release behavior of heparin from PVC films containing heparin-immobilized hydrogels. Too much immobilized heparin in PVC is not well sustained, however, and suitable amounts

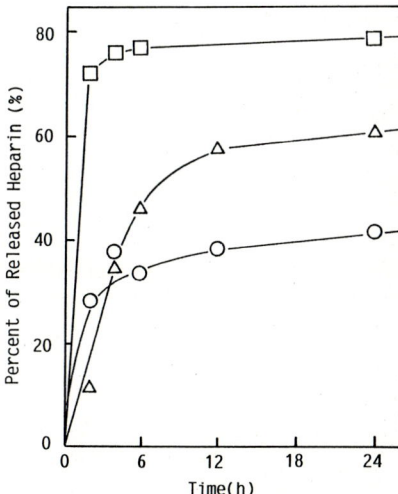

Fig. 33. Release behavior of heparin from PVC with heparinimmobilized hydrogels at 37 °C in a physiological saline. (○) Run No. 1 in Table 19, (△) Run No. 2, (□) Run No. 3

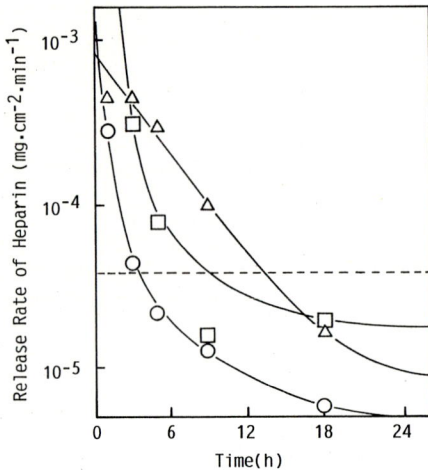

Fig. 24. Release rate of heparin from PVC with heparin-immobilized hydrogels at 37 °C in a physiological saline. (○) Run No. 1, (△) Run No. 2, (□) Run No. 3

of heparin were effectively immobilized and released slowly from the PVC matrix. A lower concentration of heparin is sustained in hydrogel matrix for a longer period. In this system, hydrogel is considered as a reservoir for the hydrophilic biologically active molecules. As the heparin comes out of the reservoir into the hydrophobic PVC matrix it moves quickly through the hydrophobic domain. This system is similar to hydrophobic drugs' sustain-release system using polyether/poly(urethane-urea) reported by Yui et al. [83], though the hydrophobic-hydrophilic combination is reversed.

It has been reported that a heparin release rate of at least 4×10^{-5} mg/cm^2/min is necessary to avoid thrombus formation [87]. Shown in Fig. 24 is the rate of heparin release from PVC at 37 °C in a physiological saline. In case of Sample No. 2, the release rate is higher than this guideline for a 12 h period which suggests that PVC having heparin-immobilized hydrogels has antithrombogenicity because of heparin release from PVC matrix. Therefore, since a part of PVC surface is covered by hydrogel as shown in scanning electron micrograph, antithrombogenicity of PVC surface may be achieved in lower heparin release rate than the guideline.

4.3 Antithrombogenicity of Heparin- and/or PGI$_2$-Immobilized PVC

The amount of heparin release from PVC films was evaluated by the activated partial thromboplastin time (aPTT) test and the results are summarized in Table 20. aPTT was measured by a modified method of Proctor and Rapaport [88], using rabbit brain phospholipid, Patorontin (Hoechst), and micronized kaoline suspension (Hoechst) as reagents and fibrometer (BBL), in PVC coated cups with or without heparin. In a cup coated with PVC containing hydrogels without heparin (Sample No. 8), thrombus formation took place rapidly. The PVC's containing heparin-immobilized gels prevented thrombus formation for 24 h except Sample No. 5 at the first stage. Because *N,N'*-dimethylpropyl-

New Aspects of Polymer Drugs 141

Table 20. APTT of PVC with heparin-immobilized hydrogels*

Run No.	Hydrogels	Component of Gel	Heparin (µg)	aPTT(s) 0 h	24 h	50 h
1	Sample No. 4	AAm	20.4	>60	57.6	32.7
2	Sample No. 5	AAm-DMAPAA	26.0	41.7	>60	37.8
3	Sample No. 6	NMAA	11.1	>60	>60	33.7
4	Sample No. 7	DMAA	27.7	>60	>60	31.2
5	Sample No. 8	AAm	none	26.2	39.2	29.0

* Hydrogel/(Hydrogel + PVC) = 10 wt%

aminoacrylamide (DMAPAA) works as a cationic adsorption site for anionic heparin, the heparin in Sample 5 was sustained too strongly and not enough was released for antithrombogenicity.

Biomaterial with immobilized prostaglandins show excellent antithrombotic effects by inhibition platelet aggregation [77–81]. However, some of the prostaglandins are too unstable to use. In our study, a stable and biologically active prostacyclin derivative TRK-100 (shown in Scheme 10) was used.

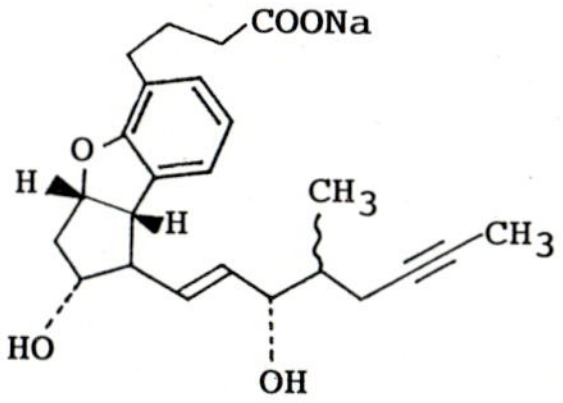

Scheme 10

Typical platelet aggregation curves for Platelet-Rich Plasma (PRP) from human blood exposed to PVC films are shown in Fig. 25. PRP was prepared as follows: Nine volumes of venous blood collected from healthy volunteers with siliconized needles and plastic syringes were added to one volume of 3.8% sodium citrate in plastic tubes. The samples were centrifuged for 10 min at 800 rpm at room temperature. After collecting the PRP, the samples were further centrifuged at 4,500 rpm for 10 min to obtain platelet-poor plasma. Platelet aggregation was measured by the turbidimetric assay [89] using collagen (2 µg/ml, Niko Bioscience, Inc) as an inducer in the PVC-treated cuvettes with or without heparin and/or PGI_2 by means of aggregometer. Percentage of aggregation was converted to percentage of inhibition of aggregation using Eq. (3) [80]:

$$\text{Degree of Inhibition (\%)} = [1 - (A_s - A_{min})/(A_{max} - A_{min})] \times 100 \qquad (3)$$

where A_s is the measured percent aggregation of the test sample, A_{min} is the minimum percentage of aggregation associated with excessive levels of TRK-100

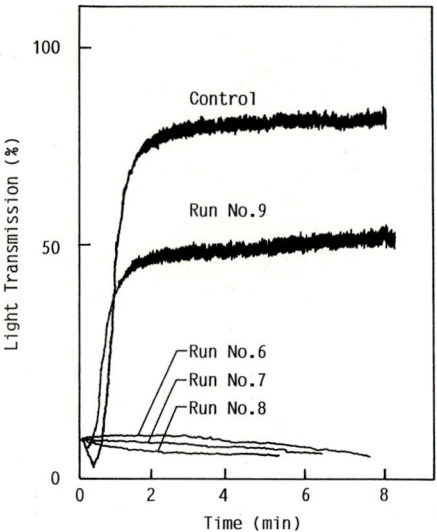

Fig. 25. Platelet aggregation curves for PRP exposed to PVC with PGI_2-immobilized hydrogels

in the PRP and A_{max} is the maximum aggregation of PVC hydrogels without TRK-100. Listed in Table 21 are the inhibition of platelet aggregation results for PVC's containing TRK-100-immobilized hydrogels calculated by Eq. (1) after standing for 24 h and 46 h in a physiological saline. The PVC surface containing AAm gel did not inhibit platelet aggregation, although it is better than glass. Regardless of the kind of hydrogel used, 100% inhibition of platelet aggregation was observed for all the PVC's containing TRK-100-immobilized hydrogels for over 46 h. In vitro the inhibition of platelet aggregation for TRK-100 is half of that of PGI_2 and eight times better than PGE_1. The concentration of TRK-100 released from the PVC's containing TRK-100-immobilized hydrogels could be determined. However, a sufficient amount of TRK-100 appears to be released from the PVC films to inhibit platelet aggregation even after 46 h.

The first approach taken by Kim et al. [90] was to combine heparin and prostaglandin E_1 (PGE_1) together in a polymer matrix. Subsequently, a covalently

Table 21. Inhibition of platelet aggregation of PVC with PGI_2-immobilized hydrogel*

Run No.	Hydrogels	Component of Gel	Prostaglandin (µg)	Degree of Inhibition (%)		
				0 h	22 h	46 h
6	Sample No. 9	AAm	1.04	100	100	100
7	Sample No. 10	NMAA	0.90	100	100	100
8	Sample No. 11	DMAA	0.80	100	100	100
9	Sample No. 8	AAm	none	0	0	0

* Hydrogel/(Hydrogel + PVC) = 10 wt%

Table 22. Whole blood clotting time of PVC with heparin- and/or PGI$_2$-immobilized hydrogels[a]

Run No.	Hydrogels	Heparin (µg)	Prostaglandin (µg)	Whole blood clotting time (min)		
				0 h	22 h	46 h
10	Sample No. 4	19.1	none	>30	>45	>45
11	Sample No. 9	none	1.04	9.5	7.5	9
12[b]	Sample No. 4 + Sample No. 9	23.3	0.26	>30	>45	>45
13[c]	Sample No. 4 + Sample No. 9	6.76	0.30	>30	>45	>45
14	Sample No. 8	none	none	10	12	14

[a] Hydrogel/(Hydrogel + PVC) = 10 wt%.
[b] Sample No. 4/Sample No. 9 = 4/1 (wt/wt).
[c] Sample No. 4/Sample No. 9 = 1/1 (wt/wt).

bound conjugate of heparin and PGE$_1$ was synthesized for in vitro bioactivity studies [91]. In our study, both heparin and PGI$_2$ were physically immobilized in a PVC matrix to evaluate the antithrombogenic activity. Table 22 shows the whole blood clotting time for PVC containing heparin- and/or PGI$_2$-immobilized hydrogels after standing for 22 h and 46 h in a physiological saline. The whole blood clotting time was measured using venous blood. It was with drawn from a healthy volunteers by a clean venipuncture into a plastic syringe without frothing. After the needle was removed, 2 ml of aliquots of the whole blood were delivered immediately into 7×50 mm glass tubes coated with PVC with or without heparin and/or PGI$_2$. The tubes were gently tilted every 30 sec until a clot was observed in the tubes; the stopwatch was stopped at the point. In Table 22 is shown human whole blood clotted in a short time in a test tube coated by PVC containing acrylamide hydrogels only (Sample 8). Heparin-immobilized PVC's coated test tubes, however, were found to exhibit excellent antithrombogenicity either in the presence or absence of PGI$_2$. The results suggest that released heparin plays a more important role than anything else in antithrombogenicity of PVC for human whole blood.

In conclusion, although the blood compatibility of PVC itself is not sufficient, PVC containing heparin-immobilized hydrogels do provide high antithrombogenicity and can be controlled by changing the gel component. It should be possible to use this material for a considerable length of time. Moreover, PVC as a blood compatible material has a serious effect on platelet aggregation, whereas PVC containing PGI$_2$ immobilized hydrogels inhibits it perfectly. Therefore, PVC containing both heparin and PGI$_2$ is considered to be an excellent antithrombogenic material since it can inhibit two independent blood coagulation mechanisms without any interference with the biological activity of either heparin or PGI$_2$.

Acknowledgements: The authors wish to thank their colleagues Prof. N. Miyauchi, Dr. E. Yashima, from Faculty of Engineering, Kagoshima University, Dr. I. Maruyama from School of Medicine, Kagoshima University,

Dr. N. Morita from School, Medicine of Fukuoka University, Dr. T. Sato from Faculty of Engineering, Nagasaki University, and Prof. J. Sunamoto from Faculty of Engineering, Kyoto University for their cooperation and helpful suggestions.

5 References

1. Donaruma LG, Vogl O (eds) (1978) Polymeric drugs. Academic, New York
2. Ringsdorf H (1975) J. Polym. Sci. Polym. Symp. 51: 135
3. Sunamoto J et al. (1987) International Symposium on Polymer Drugs and Polymeric Drug Carriers, 28–30 Oct 1987, Nagasaki
4. Akashi M, Beppu K, Kikuchi I, Miyauchi N (1986) J. Macromol. Sci. Chem. A23: 1233
5. Akashi M, Tanaka Y, Miyazaki T, Miyauchi N (1987) J. Bioactive and Compatible Polym. 2: 120
6. Akashi M, Wada M, Yanase S, Miyauchi N (1989) J. Polym. Sci. Polym. Lett. Ed. 27: 377
7. Yashima E, Uchida S, Akashi M, Miyauchi N, Morita N, Minota T (1990) J. Bioactive and Compatible Polym. 5: 53
8. Akashi M, Miyauchi N, Morita N, Minota T (1987) J. Bioactive and Compatible Polym. 2: 232
9. Akashi M, Tanaka Y, Iwasaki H, Miyauchi N (1987) Angew. Makromol. Chem. 147: 207
10. Akashi M, Sakagami E, Tajima T, Yashima E, Miyauchi N (1988) Nucleic Acids Res. Symp. Ser. 2: 61
11. Akashi M, Iwasaki H, Miyauchi N, Sato T, Sunamoto J, Takemoto K (1989) J. Bioactive and Compatible Polym. 4: 126
12. Akashi M. Takeda S, Miyazaki T, Yashima E, Miyauchi N, Maruyama I, Okadome T, Murata Y (1989) J. Bioactive and Compatible Polym. 4: 4
13. Takemoto K (1976) J. Polym. Sci. Polym. Symp. 55: 105
14. Takemoto K (1978): In: Donaruma LG, Vogl O (eds) Polymeric drugs. Academic, New York, p 103
15. Takemoto K, Inaki Y (1981) Adv. Polym. Sci. 41: 1
16. Takemoto K (1985) Makromol. Chem. Suppl. 12: 293
17. Inaki Y, Takemoto K (1985) Makromol. Chem. Suppl. 12: 91
18. Takemoto K, Inaki Y (1988) Acta Polymer. 39: 33
19. Pitha J, Akashi M, Draminski M (1980) In: Goldberg E.P, Nakajima A, (eds) Biomedical polymer; polymeric material and pharmaceuticals for biomedical use. Academic, New York, p 271
20. Tada M, Chem. Lett. 1975: 129
21. Ozaki S, Ike 1, Mizuno H, Ishikawa K, Mori H (1975) Bull. Chem. Soc. Jpn. 50: 2406
22. Umrigar PP, Ohashi S, Butler GB (1979) J. Polym. Sci. Polym. Chem. Ed. 17: 351
23. Gebelein CG, Morgan RM, Glowacky R, Baig W (1981) Biomedical and Dental Applications of Polymers 191
24. Hartsough RR, Gebelein CG (1984) Polym. Mater. Sci. Eng. 51: 131
25. Buur A, Bundgaad H (1984) Int. J. Pharm. 21: 349
26. Ouchi T, Yuyama H, Vogl O (1985) Makromol. Chem. Rapid Commun. 6: 815
27. Akashi M, Kita Y, Inaki Y, Takemoto K (1977) Makromol. Chem. 178: 1211
28. Buur A, Bundgaard H (1986) J. Pharm. Sci. 75: 522
29. Takemoto K, Akashi M, Inaki Y (1974) J. Polym. Sci. Polym. Chem. Ed. 12: 1861
30. Akashi M, Inaki Y, Takemoto K (1977) Makromol. Chem. 178: 353
31. Akashi M, Kita Y, Inaki Y, Takemoto K (1979) J. Polym. Sci. Polym. Chem. Ed. 17: 301

32. Akashi M, Yamashita I, Miyauchi N (1984) Angew. Makromol. Chem. 122: 147
33. Akashi M, Kirikihira I, Miyauchi N (1985) Angew. Makromol. Chem. 133: 81
34. Miyauchi N, Kirikihira I, Li X, Akashi M (1988) J. Polym. Sci. Polym. Chem. Ed. 26: 1561
35. Akashi M, Yanagi T, Yashima E, Miyauchi N (1989) J. Polym. Sci. Polym. Chem. Ed. 27: 3521
36. Akashi M, Chao D, Yashima E, Miyauchi N (1990) J. Appl. Polym. Sci. 39: 2027
37. Jones JA, Ottenbrite RM (1986) J. Polym. Sci. Polym. Chem. Ed. 24: 1487
38. Ottenbrite RM, Butler GB (1984) In: Anticancer and interferon agents. Dekker, New York, p 247
39. Morawetz H, Song WR (1966) J. Am. Chem. Soc. 88: 5714
40. Ikada Y, Tabata Y (1986) J. Bioactive and Compatible Polym. 1: 32
41. Maeda H, Ueda M, Morinaga T, Matsumoto T (1985) J. Med. Chem. 28: 455
42. Konno T, Maeda H, Iwai K, Tashiro S, Miki S, Morinaga T, Morinaga M, Hiraoka T, Yokoyama I (1983) Eur. J. Cancer Clin. Oncol. 19: 1053
43. Donaruma LG, Ottenbrite RM, Vogl O (1980) Anionic polymeric drugs. Wiley, New York
44. Kaetsu I, Yoshida M, Yamada A (1980) J. Biomed. Mater. Res. 14: 185
45. Kotaka S, Tsutsumiuchi K, Watanabe I (1981) J. Jpn. Soc. Cancer Ther. 16: 460
46. Yamashita R, Saketoku M, Hirano M, Iwa T (1985) Jinkozoki 14: 818
47. Ozaki S, Ike Y, Mizuno H, Ishikawa K, Mori H (1977) Bull. Chem. Soc. Jpn. 50: 2406
48. Ozaki S, Watanabe T, Hoshiko T, Mizuno H, Ishikawa K, Mori H, (1984) Chem. Pharm. Bull. 32: 733
49. Hoshiko T, Ozaki S, Watanabe Y, Ogasawara T, Yamauchi S, Fujiwara K, Hoshi A, Iigo M (1985) Chem. Pharm. Bull. 33: 2832
50. Ozaki S, Watanabe Y, Hoshiko T, Nagase T, Ogasawara T, Furukawa H, Uemura A, Ishikawa K, Mori H, Hoshi A, Iigo M, Tokuzene R (1985) Chem. Pharm. Bull. 34: 150
51. Ozaki S, Nagase T, Tamai H, Mori H, Hoshi A, Iigo M (1987) Chem. Pharm. Bull. 35: 3894
52. Ahmad S, Ozaki S, Nagase T, Iigo M, Tokuzene R, Hoshi A (1987) Chem. Pharm. Bull. 35: 4137
53. Kametani T (1980) J. Med. Chem. 23: 1324
54. Ouchi T, Fujie, Jokei S, Sakamoto Y, Chikashita H, Inoi T, Vogl O (1986) J. Polym. Sci. Polym. Chem. Ed. 24: 2059
55. Ouchi T, Yuyama H (1987) J. Polym. Sci. Polym. Lett. Ed. 25: 279
56. Pryzblski M, Fell E, Ringsdorf H (1978) Makromol. Chem. 179: 1719
57. Kaye H (1970) J. Am. Chem. Soc. 92: 5777
58. Pitha J, Pitha PM, Ts'o POP (1970) Biochem. Biophys. Acta. 204: 39
59. Kondo K, Iwasaki H, Nakatani K, Ueda N, Takemoto K, Imoto M (1969) Makromol. Chem. 125: 42
60. Tinoko IJr (1960) J. Am. Chem. Soc. 82: 4765
61. Brahms J, Maurizot JC, Michelson AM (1967) J. Mol. Biol. 25: 465
62. Chan SI, Nelson JH (1969) J. Am. Chem. Soc. 91: 168
63. Pitha J, Ts'o POP (1968) J. Org. Chem. 33: 1341
64. Akashi M, Okimoto T, Inaki Y, Takemoto K (1979) J. Polym. Sci. Polym. Chem. Ed. 17: 905
65. Stevens CL, Felsenfeld G (1964) Biopolymers. 2: 293
66. Thomas GJJr., Kyogoku Y (1967) J. Am. Chem. Soc. 89: 4170
67. Maeda H (1985) Oncologia. 15: 90
68. Sunamoto J, Iwamoto K, Takada M, Yuzuriha T, Katayama (1984) In: Anderson JM (ed) Recent advances in drug delivery system. Plenum, New York, p 153
69. Sato T, Kojima K, Ohda T, Sunamoto J, Ottenbrite RM (1986) J. Bioactive Compatible Polym. 1: 448 (1986)
70. Levy HB (1986) J. Bioactive and Compatible Polym. 1: 348
71. Bunting S, Gryglewski R, Moncado S, Vane JR (1976) Prostaglandins. 12: 897
72. Marcum JA, Fritze L, Galli SJ, Karp G, Rosenberg RD (1983) Am. J. Physiol. 245: H725 (1983)
73. Maruyama I, Bell CE, Majerus PW (1985) J. Cell. Biol. 101: 363

74. Gott VL, Whitten JD, Dutton RC (1963) Science. 142: 1297
75. Ito Y (1987) Biomaterials Appl. 2: 235
76. Senatore F, Bernath F, Meisner K (1986) J. Biomed. Mat. Res. 20: 177
77. Grode GA, Pitman J, Crowley JP, Leiniger RI, Falb RD (1974) Trans. Am. Soc. Artif. Intern. Organs. 20: 38
78. McRea JC, Kim SW (1978) Trans. Am. Soc. Artif. Intern. Organs. 24: 746
79. McRea JC, Ebert CD, Kim SW (1981) Trans. Am. Soc. Artif. Intern. Organs. 27: 511
80. Ebert CD, Lee ES., Kim SW (1982) J. Biomed. Mat. Res. 16: 629
81. Chandy T, Sharma CP (1984) J. Biomed. Mat. Res. 18: 1115
82. Hoffman AS (1986) Chemtech. 16: 426
83. Miyama H, Harumiya N, Mori Y, Tanzawa (1977) J. Biomed. Mater. Res. 11: 251
84. Tanzawa H, Mori Y, Haruyama N, Miyama H, Hori M, Oshima N, Idezuki Y (1973) Trans. Am. Soc. Artif. Intern. Organs. 19: 188
85. Cottonaro CN, Roohk HV, Bartler RH, Servas FM, Sperling DR (1982) Trans. Am. Soc. Artif. Intern. Organs. 28: 478
86. Yui N, Kataoka K, Yamada A, Sakurai Y, Sanui K, Ogata N (1986) Makromol. Chem. Rapid Commun. 7: 747
87. Idezuki Y, Watanabe H, Hagiwara M, Hanasugi K, Mori Y, Nagaoka S, Hagio M, Hamamoto, Tanzawa H (1975) Trans. Am. Soc. Artif. Intern. Organs. 21: 436
88. Proctor RR, Rapaport SI (1961) Am. J. Clin. Pathol. 36: 212
89. O'Brien JR (1962) J. Clin. Pathol. 15: 446
90. Kim SW, Ebert C, Lin J, McRea J (1983) Am. Soc. Artif. Intern. Organs. 6: 76
91. Jacobs H, Kim SW (1986) J. Pharm. Sci. 75: 172

Author Index Volumes 1–97

Allegra, G. and *Bassi, I. W.:* Isomorphism in Synthetic Macromolecular Systems. Vol. 6. pp. 549–574.
Akashi, M. and *Takemoto, K.:* New Aspects of Polymer Drugs. Vol. 97, pp. 107–146.
Andrade, J. D., Hlady, V.: Protein Adsorption and Materials Biocompability: A. Tutorial Review and Suggested Hypothesis. Vol. 79, pp. 1–63.
Andreis, M. and *Koenig, J. L.:* Application of NMR to Crosslinked Polymer Systems. Vol. 89, pp. 69–160.
Andrews, E. H.: Molecular Fracture in Polymers. Vol. 27, pp. 1–66.
Anufrieva, E. V. and *Gotlib, Yu. Ya.:* Investigation of Polymers in Solution by Polarized Luminescence. Vol. 40, pp. 1–68.
Apicella, A. and *Nicolais, L.:* Effect of Water on the Properties of Epoxy Matrix and Composite. Vol. 72, pp. 69–78.
Apicella, A., Nicolais, L. and *de Cataldis, C.:* Characterization of the Morphological Fine Structure of Commercial Thermosetting Resins Through Hygrothermal Experiments. Vol. 66, pp. 189–208.
Argon, A. S., Cohen, R. E., Cebizlioglu, O. S. and *Schwier, C.:* Crazing in Block Copolymers and Blends. Vol. 52/53, pp. 275–334.
Argon, A. S. and *Cohen, R. E.:* Crazing and Toughness of Block Copolymers and Blends. Vol. 91/92, pp. 301–352.
Aronhime, M. T., Gillham, J. K.: Time-Temperature Transformation (TTT) Cure Diagram of Thermosetting Polymeric Systems. Vol. 78, pp. 81–112.
Arridge, R. C. and *Barham, P. J.:* Polymer Elasticity. Discrete and Continuum Models. Vol. 46, pp. 67–117.
Aseeva, R. M., Zaikov, G. E.: Flammability of Polymeric Materials. Vol. 70, pp. 171–230.
Ayrey, G.: The Use of Isotopes in Polymer Analysis. Vol. 6, pp. 128–148.

Bässler, H.: Photopolymerization of Diacetylenes. Vol. 63, pp. 1–48.
Baldwin, R. L.: Sedimentation of High Polymers. Vol. 1, pp. 451–511.
Bascom, W. D.: The Wettability of Polymer Surfaces and the Spreading of Polymer Liquids. Vol. 85, pp. 89–124.
Belta-Calleja, F. J.: Microhardness Relating to Crystalline Polymers. Vol. 66, pp. 117–148.
Barbé, P. C., Cecchin, G. and *Noristi, L.:* The Catalytic System Ti-Complex/$MgCl_2$. Vol. 81, pp. 1–83.
Barton, J. M.: The Application of Differential Scanning Calorimetry (DSC) to the Study of Epoxy Resins Curing Reactions. Vol. 72, pp. 111–154.
Ballauff, M. and *Wolf, B. A.:* Thermodynamically Induced Shear Degradation. Vol. 84, pp. 1–31.
Basedow, A. M. and *Ebert, K.:* Ultrasonic Degradation of Polymers in Solution. Vol. 22, pp. 83–148.
Batz, H.-G.: Polymeric Drugs. Vol. 23, pp. 25–53.
Baur, H. see Wunderlich, B.: Vol. 87, pp. 1–121.
Bell, J. P. see Schmidt, R. G.: Vol. 75, pp. 33–72.

Bekturov, E. A. and *Bimendina, L. A.:* Interpolymer Complexes. Vol. 41, pp. 99–147.
Berger, L. L. see Kramer, E. J.: Vol. 91/92, pp. 1–68.
Bergsma, F. and *Kruissink, Ch. A.:* Ion-Exchange Membranes. Vol. 2, pp. 307–362.
Berlin, Al. Al., Volfson, S. A., and *Enikolopian, N. S.:* Kinetic of Polymerization Process. Vol. 38, pp. 89–140.
Berry, G. C. and *Fox, T. G.:* The Viscosity of Polymers and Their Concentrated Solutions. Vol. 5, pp. 261–357.
Bevington, J. C.: Isotopic Methods in Polymer Chemistry. Vol. 2, pp. 1–17.
Beylen, M. van, Bywater, S., Smets, G., Szwarc, M. and *Worsfold, D. J.:* Developments in Anionic Polymerization — A Critical Review. Vol. 86, pp. 87–143.
Bhuiyan, A. L.: Some Problems Encountered with Degradation Mechanisms of Addition Polymers. Vol. 47, pp. 1–65.
Biedermann, H. see Osada, Y.: Vol. 95, pp. 57–110.
Billingham, N. C. and *Calvert, P. D.:* Electrically Conducting Polymers — A Polymer Science Viewpoint. Vol. 90, pp. 1–104.
Bird, R. B., Warner, Jr., H. R. and *Evans, D. C.:* Kinetic Theory and Rheology of Dumbbell Suspension with Brownian motion. Vol. 8, pp. 1–90.
Biswas, M. and *Maity, C.:* Molecular Sieves as Polymerization Catalysts. Vol. 31, pp. 47–88.
Biswas, M., Packirisamy, S.: Synthetic Ion-Exchange Resins. Vol. 70, pp. 71–118.
Block, H.: The Nature and Application of Electrical Phenomena in Polymers. Vol. 33, pp. 93–167.
Bodor, G.: X-ray Line Shape Analysis. A. Means for the Characterization of Crystalline Polymers. Vol. 67, pp. 165–194.
Böhm, L. L., Chmelir̆, M., Löhr, G., Schmitt, B. J. and *Schulz, G. V.:* Zustände und Reaktionen des Carbanions bei der anionischen Polymerisation des Styrols. Vol. 9, pp. 1–45.
Bölke, P. see Hallpap, P.: Vol. 86, pp. 175–236.
Bormashenko, E. Yu. see Fridman, M. L.: Vol. 93, pp. 81–136.
Boutevin, B.: Telechelic Oligomers by Radical Reactions. Vol. 94, pp. 69–106.
Boué, F.: Transient Relaxation Mechanisms in Elongated Melts and Rubbers Investigated by Small Angle Neutron Scattering. Vol. 82, pp. 47–103.
Bovey, F. A. and *Tiers, G. V. D.:* The High Resolution Nuclear Magnetic Resonance Spectroscopy of Polymers. Vol. 3, pp. 139–195.
Braun, J.-M- and *Guilett, J. E.:* Study of Polymers by Inverse Gas Chromatography. Vol. 21, pp. 107–145.
Breitenbach, J. W., Olaj, O. F. und *Sommer, F.:* Polymerisationsanregung durch Elektrolyse. Vol. 9, pp. 47–227.
Bresler, S. E. and *Kazbekov, E. N.:* Macroradical Reactivity Studied by Electron Spin Resonance. Vol. 3, pp. 688–711.
Brosse, J.-C., Derouet, D., Epaillard, F., Soutif, J.-C., Legeay, G. and *Dušek, K.:* Hydroxyl-Terminated Polymers Obtained by Free Radical Polymerization. Synthesis, Characterization, and Application. Vol. 81, pp. 167–224.
Bucknall, C. B.: Fracture and Failure of Multiphase Polymers and Polymer Composites. Vol. 27, pp. 121–148.
Burchard, W.: Static and Dynamic Light Scattering from Branched Polymers and Biopolymers. Vol. 48, pp. 1–24.
Bywater, S.: Polymerization Initiated by Lithium and Its Compounds. Vol. 4, pp. 66–110.
Bywater, S.: Preparation and Properties of Star-branched Polymers. Vol. 30, pp. 89–116.
Bywater, S. see Beylen, M. van: Vol. 86, pp. 87–143.

Calvert, P. D. see Billingham, N. C.: Vol. 90, pp. 1–104.
Candau, S., Bastide, J. and *Delsanti, M.:* Structural. Elastic and Dynamic Properties of Swollen Polymer Networks. Vol. 44, pp. 27–72.
Carrick, W. L.: The Mechanism of Olefin Polymerization by Ziegler-Natta Catalysts. Vol. 12, pp. 65–86.
Casale, A. and *Porter, R. S.:* Mechanical Synthesis of Block and Graft Copolymers. Vol. 17, pp. 1–71.

Cecchin, G. see *Barbé, P. C.*: Vol. 81, pp. 1–83.
Cerf, R.: La dynamique des solutions de macromolecules dans un champ de vitresses. Vol. 1, pp. 382–450.
Cesca, S., Priola, A. and *Bruzzone, M.:* Synthesis and Modification of Polymers Containing a System of Conjugated Double Bonds. Vol. 32, pp. 1–67.
Chiellini, E., Solaro, R., Galli, G. and *Ledwith, A.:* Optically Active Synthetic Polymers Containing Pendant Carbazolyl Groups. Vol. 62, pp. 143–170.
Cicchetti, O.: Mechanisms of Oxidative Photodegradation and of UV Stabilization of Polyolefins. Vol. 7, pp. 70–112.
Clark, A. H. and *Ross-Murphy, S. B.:* Structural and Mechanical Properties of Biopolymer Gels. Vol. 83, pp. 57–193.
Clark, D. T.: ESCA Applied to Polymers. Vol. 24, pp. 125–188.
Cohen, R. E. see *Argon, A. S.*: Vol. 91/92, pp. 301–352.
Colemann, Jr., L. E. and *Meinhardt, N. A.:* Polymerization Reactions of Vinyl Ketones. Vol. 1, pp. 159–179.
Comper, W. D. and *Preston, B. N.:* Rapid Polymer Transport in Concentrated Solutions. Vol. 55, pp. 105–152.
Corner, T.: Free Radical Polymerization – The Synthesis of Graft Copolymers. Vol. 62, pp. 95–142.
Cresceni, V.: Some Recent Studies of Polyelectrolyte Solutions. Vol. 5, pp. 358–386.
Crivello, J. V.: Cationic Polymerization – Iodonium and Sulfonium Salt Photoinitiators, Vol. 62, pp. 1–48.

Dave, R. see *Kardos, J. L.*: Vol. 80, pp. 101–123.
Davydov, B. E. and *Krentsel, B. A.:* Progress in the Chemistry of Polyconjugated Systems. Vol. 25, pp. 1–46.
Derouet, F. see *Brosse, J.-C.*: Vol. 81, pp. 167–224.
Dettenmaier, M.: Intrinsic Crazes in Polycarbonate Phenomenology and Molecular Interpretation of a New Phenomenon. Vol. 52/53, pp. 57–104.
Dettenmaier, M. and *Leberger, D.:* Crazing of Polymer-Diluent Mixtures. Vol. 91/92, pp. 119–136.
Diaz, A. F., Rubinson, J. F. and *Mark, H. B., Jr.:* Electrochemistry and electrode Application of Electroactive/Conductive Polymers. Vol. 84, pp. 113–140.
Dobb, M. G. and *McIntyre, J. E.:* Properties and Applications of Liquid-Crystalline Main-Chain Polymers. Vol. 60/61, pp. 61–98.
Döll, W.: Optical Interference Measurements and Fracture Mechanics Analysis of Crack Tip Craze Zones. Vol. 52/53, pp. 105–168.
Döll, W. and *Könczöl, L.:* Micromechanics of Fracture under Static and Fatigue Loading: Optical Interferometry of Drack Tip Craze Zones. Vol. 91/92, pp. 137–214.
Doi, Y. see *Keli, T.*: Vol. 73/74, pp. 201–248.
Dole, M.: Calorimetric of States and Transitions in Solid High Polymers. Vol. 2, pp. 221–274.
Donnet, J. B., Vidal, A.: Carbon Black-Surface Properties and Interactions with Elastomers. Vol. 76, pp. 103–128.
Dorn, K., Hupfer, B. and *Ringsdorf, H.:* Polymeric Monolayers and Liposomes as Models for Biomembranes How to Bridge the Gap Between Polymer Science and Membrane Biology? Vol. 64, pp. 1–54.
Dreyfuss, P. and *Dreyfuss, M. P.:* Polytetrahydrofuran. Vol. 4, pp. 528–590.
Drobnik, J. and *Rypáček, F.:* Soluble Synthetic Polymers in Biological Systems. Vol. 57, pp. 1–50.
Dröscher, M.: Solid State Extrusion of Semicrystalline Copolymers. Vol. 47, pp. 120–138.
Duduković, M. P. see *Kardos, J. L.*: Vol. 80, pp. 101–123.
Drzal, L. T.: The Interphase in Epoxy Composites. Vol. 75, pp. 1–32.
Dušek, K.: Network Formation in Curing of Epoxy Resins. Vol. 78, pp. 1–58.
Dušek, K. and *Prins, W.:* Structure and Elasticity of Non-Crystalline Polymer Networks. Vol. 6, pp. 1–102.

Dušek, K. see Brosse, J.-C.: Vol. 81, pp. 167–224.
Duncan, R. and *Kopeček, J.:* Soluble Synthetic Polymers as Potential Drug Carriers. Vol. 57, pp. 51–101.

Eastam, A. M.: Some Aspects of the Polymerization of Cyclic Ethers. Vol. 2, pp. 18–50.
Ehrlich, P. and *Mortimer, G. A.:* Fundamentals of the Free-Radical Polymerization of Ethylene. Vol. 7, pp. 386–448.
Eisenberg, A.: Ionic Forces in Polymers. Vol. 5, pp. 59–112.
Eiss, N. S. Jr. see Yorkgitis, E. M.: Vol. 72, pp. 79–110.
Elias, H.-G., Bareiss, R. und *Watterson, J. G.:* Mittelwerte des Molekulargewichts und anderer Eigenschaften. Vol. 11, pp. 111–204.
Elsner, G., Riekel, Ch. and *Zachmann, H. G.:* Synchrotron Radiation Physics. Vol. 67, pp. 1–58.
Elyashevich, G. K.: Thermodynamics and Kinetics of Orientational Crystallization of Flexible-Chain Polymers. Vol. 43, pp. 207–246.
Enikolopyan, N. S., Friedman, M. L., Stalnova, I. O. and *Popov, V. L.:* Filled Polymers: Mechanical Properties and Processability. Vol. 96, pp. 1–67.
Enikolopyan, N. S. see Ponomarenko, A. T.: Vol. 96, pp. 125–147.
Enkelmann, V.: Structural Aspects of the Topochemical Polymerization of Diacetylenes. Vol. 63, pp. 91–136.
Entelis, S. G., Evreinov, V. V., Gorshkov, A. V.: Functionally and Molecular Weight Distribution of Telchelic Polymers. Vol. 76, pp. 129–175.
Epaillard, F. see Brosse, J.-C.: Vol. 81, pp. 167–224.
Evreinov, V. V. see Entelis, S. G.: Vol. 76, pp. 129–175.

Ferruti, P. and *Barbucci, R.:* Linear Amino Polymers. Synthesis, Protonation and Complex Formation. Vol. 58, pp. 59–92.
Finkelmann, H. and *Rehage, G.:* Liquid Crystal Side-Chain Polymers. Vol. 60/61, pp. 99–172.
Fischer, H.: Freie Radikale während der Polymerisation, nachgewiesen und identifiziert durch Elektronenspinresonanz. Vol. 5, pp. 463–530.
Flory, P. J.: Molecular Theory of Liquid Crystals. Vol. 59, pp. 1–36.
Ford, W. T. and *Tomoi, M.:* Polymer-Supported Phase Transfer Catalyst Reaction Mechanisms. Vol. 55, pp. 49–104.
Fradet, A. and *Maréchal, E.:* Kinetics and Mechanisms of Polyesterifications. I. Reactions of Diols with Diacids. Vol. 43, pp. 51–144.
Franta, E. see Rempp, P.: Vol. 86, pp. 145–173.
Franz, G.: Polysaccharides in Pharmacy. Vol. 76, pp. 1–30.
Fridman, M. L. and *Servuk, V. D.:* Extension of Molten Polymers. Vol. 93, pp. 1–40.
Fridman, M. L. and *Peshkovsky, S. L.:* Molding of Polymers under Conditions of Vibration Effects. Vol. 93, pp. 41–80.
Fridman, M. L., Petrosyan, A.-Z., Levin, V. S. and *Bormashenko, E. Yu.:* Fundamentals of Low-Pressure Moulding of Polymer Pastes(Plastisols) and Thermoplastic Materials. Vol. 93, pp. 81–136.
Fridman, M. L. see Tunkel, V. I.: Vol. 93, pp. 137–174.
Fridman, M. L. see Sabsai, O. Yu.: Vol. 96, pp. 99–123.
Fridman, M. L. see Enikolopyan, N. S.: Vol. 96, pp. 1–67.
Friedrich, K.: Crazes and Shear Bands in Semi-Crystalline Thermoplastics. Vol. 52/53, pp. 225–274.
Fujita, H.: Diffusion on Polymer-Diluent Systems. Vol. 3, pp. 1–47.
Funke, W.: Über die Strukturaufklärung vernetzter Makromoleküle, insbesondere vernetzter Polyesterharze, mit chemischen Methoden. Vol. 4, pp. 157–235.
Furukowa, H. see Kamon, T.: Vol. 80, pp. 173–202.

Gal'braikh, L. S. and *Rigovin, Z. A.:* Chemical Transformation of Cellulose. Vol. 14, pp. 87 bis 130.
Galli, G. see Chiellini, E.: Vol. 62, pp. 143–170.

Gallot, B. R. M.: Preparation and Study of Block Copolymers with Ordered Structures, Vol. 29, pp. 85–156.
Gandini, A.: The Behaviour of Furan Derivatives in Polymerization Reactions. Vol. 25, pp. 47–96.
Gandini, A. and *Cheradame, H.:* Cationic Polymerization. Initiation with Alkenyl Monomers. Vol. 34/35, pp. 1–289.
Geckeler, K., Pillai, V. N. R., and *Mutter, M.:* Application of Soluble Polymeric Supports. Vol. 39, pp. 65–94.
Gerrens, H.: Kinetik der Emulsionspolymerisation. Vol. 1, pp. 234–328.
Ghiggino, K. P., Roberts, A. J. and *Phillips, D.:* Time-Resolved Fluorescence Techniques in Polymer and Biopolymer Studies. Vol. 40, pp. 69–167.
Gilham, J. K. see Aronhime, M. T.: Vol. 78, pp. 81–112.
Glöckner, G.: Analysis of Compositional and Structural Heterogeneitis of Polymer by Non-Exclusion HPCL. Vol. 79, pp. 159–214.
Godovsky, Y. K.: Thermomechanics of Polymers. Vol. 76, pp. 31–102.
Godovsky, Yu. K. and *Papkov, V. S.:* Thermotropic Mesophases in Element-Organic Polymers. Vol. 88, pp. 129–180.
Goethals, E. J.: The Formation of Cyclic Oligomers in the Cationic Polymerization of Heterocycles. Vol. 23, pp. 103–130.
Gorshkov, A. V. see Entelis, S. G.: Vol. 76, pp. 129–175.
Gräger, H. see Kulicke, W.-M.: Vol. 89, pp. 1–68.
Graessley, W. W.: The Etanglement Concept in Polymer Rheology. Vol. 16, pp. 1–179.
Graessley, W. W.: Entagled Linear, Branched and Network Polymer Systems. Molecular Theories. Vol. 47, pp. 67–117.
Grebowicz, J. see Wunderlich, B.: Vol. 60/61, pp. 1–60.
Grebowicz, J. see Wunderlich, B.: Vol. 87, pp. 1–121.
Greschner, G. S.: Phase Distribution Chromatography. Possibilities and Limitations. Vol. 73/74, pp. 1–62.

Hagihara, V., Sonogahira, K. and *Takuhashi, S.:* Linear Polymers Containing Transition Metals in the Main. Vol. 41, pp. 149–179.
Hall, H. K. see Li, T.: Vol. 97, pp. 1–39.
Hallpap, P., Bölke, M., and *Heublein, G.:* Elucidation of Cationic Polymerization Mechanism by Means of Quantum Chemical Methods. Vol. 86, pp. 175–236.
Hasegawa, M.: Four-Center Photopolymerization in the Crystalline State. Vol. 42, pp. 1–49.
Hatano, M.: Induced Circular Dichroism in Biopolymer-Dye System. Vol. 77, pp. 1–121.
Hay, A. S.: Aromatic Polyethers. Vol. 4, pp. 496–527.
Hara, M. see Sauer, J. A.: Vol. 91/92, pp. 69–118.
Hayakowa, R. and *Wada, Y.:* Piezoelectricity and Related Properties of Polymer Films. Vol. 11, pp. 1–55.
Heidemann, E. and *Roth, W.:* Synthesis and Investigation of Collagen Model Peptides. Vol. 53, pp. 145–205.
Heinrich, G., Straube, E. and *Helmis, G.:* Ruber Elasticity of Polymer Networks: Theories. Vol. 84, pp. 33–87.
Heitz, W.: Polymeric Reagents. Polymer Design, Scope, and Limitations. Vol. 23, pp. 1–23.
Helfferich, F.: Ionenaustausch. Vol. 1, pp. 329–381.
Helmis, G. see Heinrich, G.: Vol. 84, pp. 33–87.
Hendra, P. J.: Laser-Raman Spectra of Polymers. Vol. 6, pp. 151–169.
Hendrix, J.: Position Sensitive "X-ray Detectors". Vol. 67, pp. 59–98.
Henrici-Olivé, G. and *Olivé, S.:* Oligomerization of Ethylene with Soluble Transition-Metal Catalysts. pp. 496–577.
Henrici-Olivé, G. and OLIVÉ, S.: Koordinative Polymerisation an löslichen Übergangsmetall-Katalysatoren. Vol. 6, pp. 421–472.
Henrici-Olivé, G. and *Olivé, S.:* Oligomerization of Ethylene with Soluble Transition-Metal Catalysts. Vol. 15, pp. 1–30.

Henrici-Olivé, G. and *Olivé, S.:* Molecular Interactions and Macroscopic Properties of Polyacrylonitrile and Model Substances. Vol. 32, pp. 123–152.
Henrici-Olivé, G. and *Olivé, S.:* The Chemistry of Carbon Fiber Formation from Polyacrylonitrile. Vol. 51, pp. 1–60.
Hermans, Jr., J., Lohr, D. and *Ferro, D.:* Treatment of the Folding and Unfolding of Protein Molecules in Solution According to a Lattice Model. Vol. 9, pp. 229–283.
Herz, J.-E. see Rempp, P.: Vol. 86, pp. 145–173.
Heublein, G. see Halipap, P.: Vol. 86, pp. 175–236.
Higashimura, T. and *Sawamoto, M.:* Living Polymerization and Selective Dimerization: Two Extremes of the Polymer Synthesis by Cationic Polymerization. Vol. 62, pp. 49–94.
Higashimura, T. see Masuda, T.: Vol. 81, pp. 121–166.
Hlady, V. see Andrade, J. D.: Vol. 79, pp. 1–63.
Hoffmann, A. S.: Ionizing Radiation and Gas Plasma (or Glow) Discharge Treatments for Preparation of Novel Polymeric Biomaterials. Vol. 57, pp. 141–157.
Holzmüller, W.: Molecular Mobility, Deformation and Relaxation Processes in Polymers. Vol. 26, pp. 1–62.
Hori, Y. see Kashiwabara, H.: Vol. 82, pp. 141–207.
Horie, K. and *Mita, I.:* Reactions and Photodynamics in Polymer Solids. Vol. 88, pp. 77–128.
Hutchinson, J. and *Ledwith, A.:* Photoinitiation of Vinyl Polymerization by Aromatic Carbonyl Compounds. Vol. 14, pp. 49–86.

Iizuka, E.: Properties of Liquid Crystals of Polypeptides: with Stress on the Electromagnetic Orientation. Vol. 20, pp. 79–107.
Ikada, Y.: Characterization of Graft Copolymers. Vol. 29, pp. 47–84.
Ikada, Y.: Blood-Compatible Polymers. Vol. 57, pp. 103–140.
Ikada, Y. see Tabata, Y.: Vol. 94, pp. 107–142.
Imanishi, Y.: Synthese, Conformation, and Reactions of Cyclic Peptides. Vol. 20, pp. 1–77.
Inagaki, H.: Polymer Separation and Characterization by Thin-Layer Chromatography. Vol. 24, pp. 189–237.
Inoue, S.: Asymmetric Reactions of Synthetic Polypeptides. Vol. 21, pp. 77–106.
Irie, M: Photoresponsive Polymers. Vol. 94, pp. 27–68.
Ise, N.: Polymerizations under an Electric Field. Vol. 6, pp. 347–376.
Ise, N.: The Mean Activity Coefficient of Polyelectrolytes in Aqueous Solutions and Its Related Properties. Vol. 7, pp. 536–593.
Isihara, A.: Irreversible Processes in Solutions of Chain Polymers. Vol. 5, pp. 531–567.
Isihara, A.: Intramolecular Statistics of a Flexible Chain Molecule. Vol. 7, pp. 449–476.
Isihara, A. and *Guth, E.:* Theory of Dilute Macromolecular Solutions. Vol. 5, pp. 233–260.
Ishikawa, M. see Narisawa, I.: Vol. 91/92, pp. 353–392.
Iwatsuki, S.: Polymerization of Quinodimethane Compounds. Vol. 58, pp. 93–120.

Janeschitz-Kriegl, H.: Flow Birefrigence of Elastico-Viscous Polymer Systems. Vol. 6, pp. 170–318.
Jenkins, R. and *Porter, R. S.:* Unpertubed Dimensions of Stereoregular Polymers. Vol. 36, pp. 1–20.
Jenngins, B. R.: Electro-Optic Methods for Characterizing Macromolecules in Dilute Solution. Vol. 22, pp. 61–81.
Johnston, D. S.: Macrozwitterion Polymerization. Isihara, A.: Vol. 42, pp. 51–106.

Kamachi, M.: Influence of Solvent on Free Radical Polymerization of Vinyl Compounds. Vol. 38, pp. 55–87.
Kamachi, M.: ESR Studies on Radical Polymerization. Vol. 82, pp. 207–277.
Kamide, K. and *Saito, M.:* Cellulose and Cellulose Derivatives: Recent Advances in Physical Chemistry. Vol. 83, pp. 1–57.
Kamińska, W. see Penczek, P.: Vol. 97, pp. 41–60.
Kamon, T., Furukawa, H.: Curing Mechanisms and Mechanical Properties of Cured Epoxy Resins. Vol. 80, pp. 173–202.

Kaneda, A. see *Kinjo, N.*: Vol. 88, pp. 1–48.
Kaneko, M. and *Wöhrle, D.*: Polymer-Coated Electrodes: New Materials for Science and Industry. Vol. 84, pp. 141–228.
Kaneko, M. and *Yamada, A.*: Solar Energy Conversion by Functional Polymers. Vol. 55, pp. 1–48.
Kardos, J. L., Duduković, M. P., Dave, R.: Void Growth and Resin Transport During Processing of Thermosetting — Matrix Composists. Vol. 80, pp. 101–123.
Kashiwabra, H., Shimada, S., Hori, Y. and *Sakaguchi, M.*: ESR Application to Polymer Physics — Molecular Motion in Solid Matrix in which Free Radicals are Trapped. Vol. 82, pp. 141–207.
Kawabata, S. and *Kawai, H.*: Strain Energy Density Functions of Rubber Vulcanizates from Biaxial Extension. Vol. 24, pp. 89–124.
Keli, T., Doi, Y.: Synthesis of "Living" Polyolefins with Soluble Ziegler-Natta Catalysts and Application to Block Copolymerization. Vol. 73/74, pp. 201–248.
Kelly, F. N. see *LeMay, J. D.*: Vol. 78, pp. 113–148.
Kennedy, J. P. and *Chou, T.*: Poly(isobutylene-co-β-Pinene): A New Sulfur Vulcanizable, Ozone Resistant Elastomer by Cationic Isomerization Copolymerization. Vol. 21, pp. 1–39.
Kennedy, J. P. and *Delvaux, J. M.*: Synthesis, Characterization and Morphology of Poly (butadieneg-Styrene). Vol. 38, pp. 141–163.
Kennedy, J. P. and *Gillham, J. K.*: Cationic Polymerization of Olefins with Alkylaluminium Initiators. Vol. 10, pp. 1–33.
Kennedy, J. P. and *Johnson, J. E.*: The Cationic Isomerization Polymerization of 3-Methyl-1-butene and 4-Methyl-1-pentene. Vol. 19, pp. 57–95.
Kennedy, J. P. and *Langer, Jr., A. W.*: Recent Advances in Cationic Polymerization. Vol. 3, pp. 508–580.
Kennedy, J. P. and *Otsu, T.*: Polymerization with Isomerization of Monomer Preceding Propagation. Vol. 7, pp. 369–385.
Kennedy, J. P. and *Rengachary, S.*: Correlation Between Cationic Model and Polymerization Reactions of Olefins. Vol. 14, pp. 1–48.
Kennedy, J. P. and *Trivedi, P. D.*: Cationic Olefin Polymerization Using Alkyl Halide — Alkyl-Aluminium Initiator Systems. I. Reactivity Studies. II. Molecular Weight Studies. Vol. 28, pp. 83–151.
Kenndy, J. P., Chang, V. S. and *Guyoit, A.*: Carbocationic Synthesis and Characterization of Polyolefins with Si-H and Si-Cl Head Groups. Vol. 43, pp. 1–50.
Khoklov, A. R. and *Grosberg, A. Yu.*: Statistical Theory of Polymeric Lyotropic Liquid Crystals. Vol. 41, pp. 53–97.
Kinjo, N., Ogata, M., Nishi, K. and *Keneda, A.*: Epoxy Molding Compounds as Encapsulation Materials for Microelectronic Devices. Vol. 88, pp. 1–48.
Kinloch, A. J.: Mechanics and Mechanics of Fracture of Thermosetting Epoxy Polymers. Vol. 72, pp. 45–68.
Kissin, Yu. V.: Structures of Copolymers of High Olefins. Vol. 15, pp. 91–155.
Kitagawa, T. and *Miyazawa, T.*: Neutron Scattering and Normal Vibrations of Polymers. Vol. 9, pp. 335–414.
Kitamaru, R. and *Horii, F.*: NMR Approach to the Phase Structure of Linear Polyethylene. Vol. 26, pp. 139–180.
Klosinski, P., Penczek, S.: Teichoic Acids and Their Models: Membrane Biopolymers with Polphosphate Backbones. Synthesis, Structure and Properties. Vol. 79, pp. 139–157.
Kloosterboer, J. G.: Network Formation by Chain Crosslinking Photopolymerization and its Applications in Electronics. Vol. 84, pp. 1–62.
Knappe, W.: Wärmeleitung in Polymeren. Vol. 7, pp. 477–535.
Koenig, J. L. see *Mertzel, E.* Vol. 75, pp. 73–112.
Koenig, J. L.: Fourier Transforms Infrared Spectroscopy of Polymers. Vol. 54, pp. 87–154.
Koenig, J. L. see *Andreis, M.*: Vol. 89, pp. 69–160.
Könczöl, L. see *Döll, W.*: Vol. 91/92, pp. 137–214.
Kötter, M. see *Kulicke, W.-M.*: Vol. 89, pp. 1–68.

Kolařik, J.: Secondary Relaxations in Glassy Polymers: Hydrophilic Polymethacrylates and Polyacrylates: Vol. 46, pp. 119–161.
Kong, E. S. W.: Physical Aging in Epoxy Matrices and Composites. Vol. 80, pp. 125–171.
Koningsveld, R.: Preparative and Analytical Aspects of Polymer Fractionation. Vol. 7.
Kosyanchuk, L. F. see Lipatov, Yu. S.: Vol. 88, pp. 49–76.
Kothe, G. see Müller, K.: Vol. 95, pp. 1–56.
Kovacs, A. J.: Transition vitreuse dans les polymers amorphes. Etude phénoménologique. Vol. 3, pp. 394–507.
Krässig, H. A.: Graft Co-Polymerization of Cellulose and Its Derivates. Vol. 4, pp. III–156.
Kramer, E. J.: Microscopic and Molecular Fundamentals of Crazing. Vol. 52/53, pp. 1–56.
Kramer, E. J. and *Berger, L. L.:* Fundamental Processes of Craze Growth and Fracture. Vol. 91/92, pp. 1–68.
Kraus, G.: Reinforcement of Elastomers by Carbon Black. Vol. 8, pp. 155–237.
Kratochvila, J. see Mejzlik, J.: Vol. 81, pp. 83–120.
Kreutz, W. and *Welte, W.:* A General Theory for the Evaluation of X-Ray Diagrams of Biomembranes and Other Lamellar Systems. Vol. 30, pp. 161–225.
Krimm. S.: Infrared Spectra of High Polymers. Vol. 2, pp. 51–72.
Kuhn, W., Ramel, A., Walters, D. H., Ebner, G. and *Kuhn, H. J.:* The Production of Mechanical Energy from Different Forms of Chemical Energy with Homogeneous and Cross-Striated High Polymer Systems. Vol. 1, pp. 540–592.
Kulicke, W.-M., Kötter, M. and *Gräger, H.:* Drag Reduction Phenomenon with Special Emphasis on Homogeneous Polymer Solutions. Vol. 89, pp. 1–68.
Kunitake, T. and *Okahata, Y.:* Catalytic Hydrolysis by Synthetic Polymers. Vol. 20, pp. 159 to 221.
Kurata, M. and *Stockmayer, W. H.:* Intrinsic Viscosities and Unperturbed Dimensions of Long Chain Molecules. Vol. 3, pp. 196–312.
Kurijama, I. see Nakase, Y.: Vol. 65, pp. 79–134.
Kurimura, Y.: Macromolecule-Metal Complexes – Reactions and Molecular Recognition. Vol. 90, pp. 105–138.

Lazar, M., Rychly, I., and *Rado, R.:* Crosslinking of Polyolefins. Vol. 95, pp. 149–198.
Leberger, D. see Dettenmaier, M.: Vol. 65, pp. 79–134.
Ledwith, A. and *Sherrington, D. C.:* Stable Organic Cation Salts: Ion Pair Equilibria and Use in Cationic Polymerization. Vol. 19, pp. 1–56.
Ledwith, A. see Chiellini, E.: Vol. 62, pp. 143–170.
Lee, C.-D. S. and *Daly, W. H.:* Mercaptan-Containing Polymers. Vol. 15, pp. 61–90.
Lee, C. see Li, T.: Vol. 97, pp. 1–39.
Legeay, G. see Brosse, J.-C.: Vol. 81, pp. 167–224.
LeMay, J. D., Kelly, F. N.: Structure and Ultimate Properties of Epoxy Resins. Vol. 78, pp. 113–148.
Lesná, M. see Mejzlik, J.: Vol. 81, pp. 83–120.
Levin, V. S. see Fridman, M. L.: Vol. 93, pp. 81–136.
Li, T., Lee, C. and *Hall, H. K.:* The Role of Tetramethylene Diradicals in Photo-Induced "Charge-Transfer" Cycloadditions and Copolymerization. Vol. 97, pp. 1–39.
Lindberg, J. J. and *Hortling, B.:* Cross Polarization – Magic Angle Spinning NMR Studies of Carbohydrates and Aromatic Polymers. Vol. 66, pp. 1–22.
Lipatov, Y. S.: Relacation and Viscoelastic Properties of Heterogeneous Polymeric Compositions. Vol. 22, pp. 1–59.
Lipatov, Y. S.: The Iso-Free-Volume State and Glass Transition in Amorphous Polymers: New Development of the Theory. Vol. 26, pp. 63–104.
Lipatov, Yu. S., Lipatova, T. E. and *Kosyanchuk, L. F.:* Synthesis and Structure of Macromolecular Topological Compounds. Vol. 88, pp. 49–76.
Lipatova, T. E.: Medical Polymer Adhesives. Vol. 79, pp. 65–93.
Lipatova, T. E. see Lipatov, Yu. S.: Vol. 88, pp. 49–76.
Litmanovich, A. A. see Papisov, J. M.: Vol. 90, pp. 139–180.

Lohse, F., Zweifel, H.: Photocrosslinking of Epoxy Resins. Vol. 78, pp. 59–80.
Lustoň, J. and *Vašš, F.:* Anionic Copolymerization of Cyclic Ethers with Cyclic Anhydrides. Vol. 56, pp. 91–133.

Madec, J.-P. and *Maréchal, E.:* Kinetics and Mechanisms of Polyesterifications. II. Reactions of Diacids with Diepoxides. Vol. 71, pp. 153–228.
Malkin, A. Ya. and *Zhirkov, P. V.:* Flow of Polymerizing Liquids. Vol. 95, pp. 111–148.
Malkin, A. Ya.: Rheology of Filled Polymers. Vol. 96, pp. 69–97.
Mano, E. B. and *Coutinho, F. M. B.:* Grafting on Polyamides. Vol. 19, pp. 97–116.
Maréchal, E. see Madec, J.-P. Vol. 71, pp. 153–228.
Mark, H. B., Jr. see Diaz, A. F.: Vol. 84, pp. 113–140.
Mark, J. E.: The Use of Model Polymer Networks to Elucidate Molecular Aspects of Rubberlike Elasticity. Vol. 44, pp. 1–26.
Mark, J. E. see Queslel, J. P. Vol. 71, pp. 229–248.
Maser, F., Bode, K., Pillai, V. N. R. and *Mutter, M.:* Conformational Studies on Model Peptides. Their Contribution to Synthetic, Structural and Functional Innovations on Proteins. Vol. 65, pp. 177–214.
Masuda, T. and *Higashimura, T.:* Polyacetylenes with Substituents: Their Synthesis and Properties. Vol. 81, pp. 121–166.
McGrath, J. E. see Yilgör, I.: Vol. 86, pp. 1–86.
McGrath, J. E. see Yorkgitis, E. M. Vol. 72, pp. 79–110.
McIntyre, J. E. see Dobb, M. G. Vol. 60/61, pp. 61–98.
Meerwall v., E. D.: Self-Diffusion in Polymer Systems. Measured with Field-Gradient Spin Echo NMR Methods, Vol. 54, pp. 1–29.
Mejzlik, J., Lesná, M. and *Kratochvila, J.:* Determination of the Number of Active Centers in Ziegler-Natta Polymerizations of Olefins. Vol. 81, pp. 83–120.
Mengoli, G.: Feasibility of Polymer Film Coating Through Electrointiated Polymerization in Aqueous Medium. Vol. 33, pp. 1–31.
Mertzel, E., Koenik, J. L.: Application of FT-IR and NMR to Epoxy Resins. Vol. 75, pp. 73–112.
Meyerhoff, G.: Die viscosimetrische Molekulargewichtsbestimmung von Polymeren. Vol. 3, pp. 59–105.
Millich, F.: Rigid Rods and the Characterization of Polysocyanides. Vol. 19, pp. 117–141.
Mita, I. see Horie, K.: Vol. 88, pp. 77–128.
Möller, M.: Cross Polarization – Magie Angle Sample Spinning NMR Studies. With Respect to the Rotational Isomeric States of Saturated Chain Molecules. Vol. 66, pp. 59–80.
Möller, M. see Wunderlich, B.: Vol. 87, pp. 1–121.
Morawetz, H.: Specific Jon Binding by Polyelectrolytes. Vol. 1, pp. 1–34.
Morgan, R. J.: Structure-Property Relations of Epoxies Used as Composite Matrices. Vol. 72, pp. 1–44.
Morin, B. P., Breusova, I. P. and *Rogovin, Z. A.:* Structural and Chemical Modification of Cellulose by Graft Copolymerization. Vol. 42, pp. 139–166.
Müller, K., Kothe, G., and *Wassmer, K.-H.:* Dynamic Magnetic Resonance of Liquid crystal Polymers: Molecular Organization and Macroscopic Properties. Vol. 95, pp. 1–56.
Mulvaney, J. E., Oversberger, C. C. and *Schiller, A. M.:* Anionic Polymerization. Vol. 3, pp. 106–138.

Nakase, Y., Kurijama, I. and *Odajima, A.:* Analysis of the Fine Structure of Poly(Oxymethylene) Prepared by Radiation-Induced Polymerization in the Solid State. Vol. 65, pp. 79–134.
Narisawa, I. and *Ishikawa, M.:* Crazing in Semicrystalline Thermoplastics. Vol. 91/92, pp. 353–392.
Neuse, E.: Aromatic Polybenzimidazoles. Syntheses, Properties, and Applications. Vol. 47, pp. 1–42.

Nicolais, L. see *Apicella, A.* Vol. 72, pp. 69–78.
Nikolaeva, N. E. see *Sabsai, O. Yu.*: Vol. 96, pp. 99–123.
Nishi, K. see *Kinjo, N.*: Vol. 88, pp. 1–48.
Noristi, L. see *Barbé, P. C.*: Vol. 81, pp. 1–83.
Nuyken, O., Weidner, R.: Graft and Block Copolymers via Polymeric Azo Initiators. Vol. 73/74, pp. 145–200.

Ober, Ch. K., Jin, J.-J. and *Lenz, R. W.*: Liquid Crystal Polymers with Flexible Spacers in the Main Chain. Vol. 59, pp. 103–146.
Odajima, A. see *Nakase, Y.*: Vol. 65, pp. 79–134.
Ogata, M. see *Kinjo, N.*: Vol. 88, pp. 1–48.
Okubo, T. and *Ise, N.*: Synthetic Polyelectrolytes as Models of Nucleic Acids and Esterases. Vol. 25, pp. 135–181.
Oleinik, E. F.: Epoxy-Aromatic Amine Networks in the Classy State Structure and Properties. Vol. 80, pp. 49–99.
Osaki, K.: Viscoelastic Properties of Dilute Polymer Solutions. Vol. 12, pp. 1–64.
Osada, Y.: Conversion of Chemical Into Mechanical Energy by Synthetic Polymers (Chemomechanical Systems). Vol. 82, pp. 1–47.
Osada, Y. and *Biedermann, H.*: Plasma Chemistry of Polymers. Vol. 95, pp. 57–110.
Oster, G. and *Nishijima, Y.*: Fluorescence Methods in Polymer Science. Vol. 3, pp. 313–331.
Otsu, T. see *Sato, T.* Vol. 71, pp. 41–78.
Overberger, C. G. and *Moore, J. A.*: Ladder Polymers. Vol. 7, pp. 113–150.
Oversberger, C. C. see *Mulvaney, J. E.*: Vol. 3, pp. 106–138.

Packirisamy, S. see *Biswas, M.* Vol. 70, pp. 71–118.
Papisov, J. M. and *Litmanovich, A. A.*: Molecular „Recognition" in Interpolymer Interactions and Matrix Polymerization. Vol. 90, pp. 139–180.
Papkov, S. P.: Liquid Crystalline Order in Solutions of Rigid-Chain Polymers. Vol. 59, pp. 75–102.
Papkov, V. S. see *Godovsky, Yu. K.*: Vol. 88, pp. 129–180.
Patrat, F., Killmann, E. und *Schiebener, C.*: Die Absorption von Makromolekülen aus Lösung. Vol. 3, pp. 332–393.
Patterson, G. D.: Photon Correlation Spectroscopy of Bulk Polymers. Vol. 48, pp. 125–159.
Penczek, P. and *Kamińska, W.*: Polyfunctional Cyanate Monomers as Components of Polymer Systems. Vol. 97, pp. 41–60.
Pencek, S., Kubisa, P. and *Matyjaszewski, K.*: Cationic Ring-Opening Polymerization of Heterocyclic Monomers. Vol. 37, pp. 1–149.
Pencek, S., Kubisa, P. and *Matyjaszewski, K.*: Cationic Ring-Opening Polymerization; 2. Synthetic Applications. Vol. 68/69, pp. 1–298.
Penczek, S. see *Klosinski, P.*: Vol. 79, pp. 139-157.
Peshkovsky, S. L. see *Fridman, M. L.*: Vol. 93, pp. 41–80.
Peticolas, W. L.: Inelastic Laser Light Scattering from Biological and Synthetic Polymers. Vol. 9, pp. 285–333.
Petropoulos, J. H.: Membranes with Non-Homogeneous Sorption Properties. Vol. 64, pp. 85–134.
Petrosyan, A. Z. see *Fridmann, M. L.*: Vol. 93, pp. 81–136.
Pino, P.: Optically Active Addition Polymers. Vol. 4, pp. 393–456.
Pitha, J.: Physiological Activities of Synthetic Analogs of Polynucleotides. Vol. 50, pp. 1–16.
Platé, N. A. and *Noak, O. V.*: A Theoretical Consideration of the Kinetics and Statistics of Reactions of Functional Groups of Macromolecules. Vol. 79, pp. 95–138.
Platé, N. A. see *Shibaev, V. P.* Vol. 60/61, pp. 173–252.
Plesch, P. H.: The Propagation Rate-Constants in Cationic Polymerisations. Vol. 8, pp. 137 to 154.
Pomogailo, A. D. and *Uflyand, I. E.*: Polymers Containing Metallochelate Units. Vol. 97, pp. 61–105.

Ponomarenko, A. T., Shevchenko, V. G. and *Enikolopyan, N. S.:* Formation Processes and Properties of Conducting Polymer Composites. Vol. 96, pp. 125–147.
Popov, V. L. see Enikolopyan, N. S.: Vol. 96, pp. 1–67.
Porod, G.: Anwendung und Ergebnisse der Röntgenkleinwinkelstreuung in festen Hochpolymeren. Vol. 2, pp. 363–400.
Pospišil, J.: Transformations of Phenolic Antioxidants and the Role of Their Products in the Long-Term Properties of Polyolefins. Vol. 36, pp. 69–133.
Postelnek, W., Colemann, L. E., and *Lovelace, A. M.:* Fluorine-Containing Polymers. I. Fluorinated Vinyl Polymers with Functional Groups, Condensation Polymers, and Styrene Polymers. Vol. 1, pp. 75–113.

Queslel, J. P. and *Mark, J. E.:* Molecular Interpretation of the Moduli Elastomeric Polymer Networks of Know Structure. Vol. 65, pp. 135–176.
Queslel, J. P. and *Mark, J. E.:* Swelling Equilibrium Studies of Elastomeric Network Structures. Vol. 71, pp. 229–248.

Rado, R. see Lazar, M.: Vol. 95, pp. 149–198.
Rehage, G. see Finkelmann, H. Vol. 60/61, pp. 99–172.
Rempp, P. F. and *Franta, E.:* Macromonomers: Synthesis, Characterization and Applications. Vol. 58, pp. 1–54.
Rempp, P., Herz, J. and *Borchard, W.:* Model Networks. Vol. 26, pp. 107–137.
Rempp, P., Franta, E., and *Herz, J.-E.:* Macromolecular Engineering by Anionic Methods. Vol. 86, pp. 145–173.
Richards, R. W.: Small Angle Neutron Scattering from Block Copolymers. Vol. 71, pp. 1–40.
Rigbi, Z.: Reinforcement of Rubber by Carbon Black. Vol. 36, pp. 21–68.
Rigbi, D. see Roc, R.-J.: Vol. 82, pp. 103–141.
Roe, R.-J. and *Rigby, D.:* Phase Relations and Miscibility in Polymer Blends Containing Copolymers. Vol. 82, pp. 103–141.
Rogovin, Z. A. and *Gabrielyan, G. A.:* Chemical Modifications of Fibre Forming Polymers and Copolymers of Acrylonitrile. Vol. 25, pp. 97–134.
Roha, M.: Ionic Factors in Steric Control. Vol. 4, pp. 353–392.
Roha, M.: The Chemistry of Coordinate Polymerization of Dienes. Vol. 1, pp. 512–539.
Ross-Murphy, S. B. see Clark, A. H.: Vol. 83, pp. 57–193.
Rostami, S. see Walsh, D. J. Vol. 70, pp. 119–170.
Rozengerk, v. A.: Linetics, Thermodynamics and Mechanism of Reactions of Epoxy Oligomers with Amines. Vol. 75, pp. 113–166.
Rubinson, J. F. see Diaz, A. F.: Vol. 84, pp. 113–140.
Rychly, I. see Lazar, M.: Vol. 95, pp. 149–198.

Sabsai, O. Yu., Nikolaeva, N. E. and *Fridman, M. L.:* Rheology of Gas-Containing Polymer Systems. Vol. 96, pp. 99–123.
Safford, G. J. and *Naumann, A. W.:* Low Frequency Motions in Polymers as Measured by Neutron Inelastic Scattering. Vol. 5, pp. 1–27.
Sakaguchi, M. see Kashiwabara, H.: Vol. 82, pp. 141–207.
Saito, M. see Kamide, K.: Vol. 83, pp. 1–57.
Sato, T. and *Otsu, T.:* Formation of Living Propagating Radicals in Microspheres and Their Use in the Synthesis of Block Copolymers. Vol. 71, pp. 41–78.
Sauer, J. A. and *Chen, C. C.:* Crazing and Fatigue Behavior in One and Two Phase Glassy Polymers. Vol. 52/53, pp. 169–224.
Sauer, J. A. and *Hara, M.:* Effect of Molecular Variables on Crazing and Fatigue of Polymers. Vol. 91/92, pp. 69–118.
Sawamoto, M. see Higashimura, T. Vol. 62, pp. 49–94.
Schiller, A. M. see Mulvaney, J. E.: Vol. 3, pp. 106–138.
Schirrer, R.: Optical Interferometry: Running Crack-Tip Morphologies and Craze Material Properties. Vol. 91/92, pp. 215–262.
Schmidt, R. G., Bell, J. P.: Epoxy Adhesion to Metals. Vol. 75, pp. 33–72.

Schuerch, C.: The Chemical Synthesis and Properties of Polysaccharides of Biomedical Interest. Vol. 10, pp. 173–194.
Schulz, R. C. und *Kaiser, E.:* Synthese und Eigenschaften von optischen aktiven Polymeren. Vol. 4, pp. 236–315.
Seanor, D. A.: Charge Transfer in Polymers. Vol. 4, pp. 317–352.
Semerak, S. N. and *Frank, C. W.:* Photophysics of Excimer Formation in Aryl Vinyl Polymers, Vol. 54, pp. 31–85.
Seidl, J., Malinský, J., Dušek, K. und *Heitz, W.:* Makroporöse Styrol-Divinylbenzol-Copolymere und ihre Verwendung in der Chromatographie und zur Darstellung von Ionenaustauschen. Vol. 5, pp. 113–213.
Semjonow, V.: Schmelzviskositäten hochpolymerer Stoffe. Vol. 5, pp. 387–450.
Semlyen, J. A.: Ring-Chain Equilibria and the Conformations of Polymer Chains. Vol. 21, pp. 41–75.
Sen, A.: The Copolymerization of Carbon Monoxide with Olefins. Vol. 73/74, pp. 125–144.
Senturia, S. D., Sheppard, N. F. Jr.: Dielectric Analysis of Thermoset Cure. Vol. 80, pp. 1–47.
Sevruk, V. D. see Fridmann, M. L.: Vol. 93, pp. 1–40.
Sharkey, W. H.: Polymerizations Through the Carbon-Sulphur Double Bond. Vol. 17, pp. 73–103.
Sheppard, N. F. Jr. see Senturia, S. D.: Vol. 80, pp. 1–47.
Shevchenko, V. G. see Ponomarenko, A. T.: Vol. 96, pp. 125–147.
Shibaev, V. P. and *Platé, N. A.:* Thermotropic Liquid-Crystalline Polymers with Mesogenic Side Groups, Vol. 60/61, pp. 173–252.
Shimada, S. see Kashiwabara, H.: Vol. 82, pp. 141–207.
Shimidzu, T.: Cooperative Actions in the Nucleophile-Containing Polymers. Vol. 23, pp. 55 to 102.
Shutov, F. A.: Foamed Polymers Based on Reactive Oligomers, Vol. 39, pp. 1–64.
Shutov, F. A.: Foamed Polymers. Cellular Structure and Properties. Vol. 51, pp. 155–218.
Shutov, F. A.: Syntactic Polymer Foams. Vol. 73/74, pp. 63–124.
Siesler, H. W.: Rheo-Optical Fourier-Tranform Infrared Spectroscopy: Vibrational Spěctra and Mechanical Properties of Polymers. Vol. 65, pp. 1–78.
Silvestri, G., Gambino, S., and *Filardi, G.:* Electrochemical Production of Initiators for Polymerization Processes. Vol. 38, pp. 27–54.
Sixl, H.: Spectroscopy of the Intermediate States of the Solid State Polymerization Reaction in Diacetylene Crystals. Vol. 63, pp. 49–90.
Slichter, W. P.: The Study of High Polymers by Nuclear Magnetic Resonance. Vol. 1, pp. 35–74.
Small, P. A.: Long-Chain Branching in Polymers. Vol. 18.
Smets, G.: Block and Graft Copolymers. Vol. 2, pp. 173–220.
Smets, G.: Photochromic Phenomena in the Solid Phase. Vol. 50, pp. 17–44.
Smets, G.: see Beylen, M. van: Vol. 86, pp. 87–143.
Sohma, J. and *Sakaguchi, M.:* ESR Studies on Polymer Radicals Produced by Mechanical Destruction and Their Reactivity. Vol. 20, pp. 109–158.
Solaro, R. see Chiellini, E. Vol. 62, pp. 143–170.
Sotobayashi, H. und *Springer, J.:* Oligomere in verdünnten Lösungen. Vol. 6, pp. 473–548.
Soutif, J.-C. see Brosse, J.-C.: Vol. 81, pp. 167–224.
Sperati, C. A. and *Starkweather, Jr., H. W.:* Fluorine-Containing Polymers. II. Polytetrafluoroethylene. Vol. 2, pp. 465–495.
Spiertz, E. J. see Vollenbrock, F. A.: Vol. 84, pp. 85–112.
Spiess, H. W.: Deuteron NMR – A new Tool for Studying Chain Mobility and Orientation in Polymers. Vol. 66, pp. 23–58.
Sprung, M. M.: Recent Progress in Silicone Chemistry. I. Hydrolysis of Reactive Silane Intermediates, Vol. 2, pp. 442–464.
Stahl, E. and *Brüderle, V.:* Polymer Analysis by Thermofractography. Vol. 30, pp. 1–88.
Stalnova, I. O. see Enikolopyan, N. S.: Vol. 96, pp. 1–67.
Stannett, V. T., Koros, W. J., Paul, D. R., Lonsdale, H. K., and *Baker, R. W.:* Recent Advances in Membrane Science and Technology. Vol. 32, pp. 69–121.

Stavermann, A. J.: Properties of Phantom Networks and Real Networks. Vol. 44, pp. 73–102.
Stauffer, D., Coniglio, A. and *Adam, M.:* Gelation and Critical Phenomena. Vol. 44, pp. 103 to 158.
Stille, J. K.: Diels-Adler Polymerization. Vol. 3, pp. 48–58.
Stolka, M. and *Pai, D.:* Polymers with Photoconductive Properties. Vol. 29, pp. 1–45.
Straube, E. see Heinrich, G.: Vol. 84, pp. 33–87.
Stuhrmann, H.: Resonance Scattering in Macromolecular Structure Research. Vol. 67, pp. 123–164.
Subramanian, R. V.: Electroinitiated Polymerization on Electrode. Vol. 33, pp. 35–58.
Sumitomo, H. and *Hashimoto, K.:* Polyamides as Barrier Materials. Vol. 64, pp. 55–84.
Sumitomo, H. and *Okada, M.:* Ring-Opering Polymerization of Bicyclic Acetals, Oxalactone, and Oxalactam. Vol. 28, pp. 47–82.
Szegö, L.: Modified Polyethylene Terephthalate Fibers. Vol. 31, pp. 89–131.
Swarc, M.: Termination of Anionic Polymerization. Vol. 2, pp. 275–306.
Swarc, M.: The Kinetics and Mechanism of N-carboxy-α-amino-acid Anhydride (NCA) Polymerization to Poly-amino Acids. Vol. 4, pp. 1–65.
Szwarc, M.: Thermodynamics of Polymerization with Special Emphasis on Living Polymers. Vol. 4, pp. 457–495.
Szwarc, M.: Living Polymers and Mechanisms of Anionic Polymerization. Vol. 49, pp. 1–175.
Swarc, M. see Beylen, M. van: Vol. 86, pp. 87–143.

Tabata, Y. and *Ikada, Y.:* Phagocytosis of Polymer Microspheres by Macrophages. Vol. 94, pp. 107–142.
Takahashi, A. and *Kawaguchi, M.:* The Structure of Macromolecules Adsorbed on Interfaces. Vol. 46, pp. 1–65.
Takekoshi, T.: Polyimides. Vol. 94, pp. 1–26.
Takemori, M. T.: Competition Between Crazing and Shear Flow During Fatigue. Vol. 91/92, pp. 263–300.
Takemoto, K. and *Inaki, Y.:* Synthetic Nuclei Acid Analogs. Preparation and Interactions. Vol. 41, pp. 1–51.
Takemoto, K. see Akashi, M.: Vol. 97, pp. 107–146.
Tani, H.: Stereospecific Polymerization of Aldehydes and Epoxides. Vol. 11, pp. 57–110.
Tate, B. E.: Polymerization of Itaconic Acid and Derivatives. Vol. 5, pp. 214–232.
Tazuke, S.: Photosensitized Charge Transfer Polymerization. Vol. 6, pp. 321–346.
Teramoto, A. and *Fujita, H.:* Conformation-dependet Properties of Synthetic Polypeptides in the Helix-Coil Transition Region. Vol. 18, pp. 65–149.
Theocaris, P. S.: The Mesophase and its Influence on the Mechanical Behavior of Composites. Vol. 66, pp. 149–188.
Thomas, W. M.: Mechanismus of Acrylonitrile Polymerization. Vol. 2, pp. 401–441.
Tieke, B.: Polymerization of Butadiene and Butadiyne (Diacetylene) Derivatives in Layer Structures. Vol. 71, pp. 79–152.
Tobolsky, A. V. and *DuPré, D. B.:* Macromolecular Relaxation in the Damped Torsional Oscillator and Statistical Segment Models. Vol. 6, pp. 103–127.
Tosi, C. and *Ciampelli, F.:* Applications of Infrared Spectroscopy to Ethylene-Propylene Copolymers. Vol. 12, pp. 87–130.
Tosi, C.: Sequence Distribution in Copolymers: Numerical Tables. Vol. 5, pp. 451–462.
Tran, C. see Yorgitis, E. M. Vol. 72, pp. 79–110.
Tsuchida, E. and *Nishide, H.:* Polymer-Metal Complexes and Their Catalytic Activity. Vol. 24, pp. 1–87.
Tsuji, K.: ESR Study of Photodegradation of Polymers. Vol. 12, pp. 131–190.
Tsvetkov, V. and *Andreeva, L.:* Flow and Electric Birefringence in Rigid-Chain Polymer Solutions. Vol. 39, pp. 95–207.
Tunkel, V. I. and *Fridman, M. L.:* Granulated Thermosetting Materials (Aminoplasts) — Technology. Vol. 93, pp. 137–174.
Tuzar, Z.; Kratochvill, P., and *Bohdanecký, M.:* Dilute Solution Properties of Aliphatic Polyamides. Vol. 30, pp. 117–159.

Uematsu, I. and *Uematsu, Y.*: Polypeptide Liquid Crystals. Vol. 59, pp. 37–74.
Uflyand, I. E. see Pomogailo, A. D.: Vol. 97, pp. 61–105.

Valuev, L. I. see Platé, N. A.: Vol. 79, pp. 95–138.
Valvassori, A. and *Sartori, G.*: Present Status of the Multicomponent Copolymerization Theory. Vol. 5, pp. 28–58.
Vidal, A. see Donnet, J. B. Vol. 76, pp. 103–128.
Viovy, J. L. and *Monnerie, L.*: Fluorescence Anisotropy Technique Using Synchroton Radiation as a Powerful Means for Studying the Orientation Correlation Functions of Polymer Chains. Vol. 67, pp. 99–122.
Voigt-Martin, I.: Use of Transmission Electron Microscopy to Obtain Quantitative Information About Polymers. Vol. 67, pp. 195–218.
Vollenbrock, F. A. and *Spiertz, E. J.*: Photoresist Systems for Microlithography. Vol. 84, pp. 85–112.
Voorn, M. J.: Phase Separation in Polymer Solutions. Vol. 1, pp. 192–233.

Walsh, D. J., Rostami, S.: The Miscibility of High Polymers: The Role of Specific Interactions Vol. 70, pp. 119–170.
Wassmer, K.-H. see Müller, K.: Vol. 95, pp. 1–56.
Ward, I. M.: Determination of Molecular Orientation by Spectroscopic Techniques. Vol. 66, pp. 81–116.
Ward, I. M.: The Preparation, Structure and Prooperties of Ultra-High Modulus Flexible Polymers. Vol. 70, pp. 1–70.
Weidner, R. see Nuyken, O.: Vol. 73/74, pp. 145–200.
Werber, F. X.: Polymerization of Olefins on Supported Catalysts. Vol. 1, pp. 180–191.
Wichterle, O., Šebenda, J., and *Králiček, J.*: The Anionic Polymerization of Caprolactam. Vol. 2, pp. 578–595.
Wilkes, G. L.: The Measurement of Molecular Orientation in Polymeric Solids. Vol. 8, pp. 91–136.
Wilkes, G. L. see Yorkgitis, E. M. Vol. 72, pp. 79–110.
Williams, G.: Molecular Aspects of Multiple Dielectric Relaxation Processes in Solid Polymers. Vol. 33, pp. 59–92.
Williams, J. G.: Applications of Linear Fracture Mechanics. Vol. 27, pp. 67–120.
Wöhrle, D.: Polymere aus Nitrilen. Vol. 10, pp. 35–107.
Wöhrle, D.: Polymer Square Planar Metal Chelates for Science and Industry. Synthesis, Properties and Applications. Vol. 50, pp. 45–134.
Wöhrle, D. see Kaneko, M.: Vol. 84, pp. 141–228.
Wolf, B. A.: Zur Thermodynamik der enthalpisch und der entropisch bedingten Entmischung von Polymerlösungen. Vol. 10, pp. 109–171.
Wolf, B. A. see Ballauff, M.: Vol. 84, pp. 1–31.
Wong, C. P.: Application of Polymer in Encapsulation of Electronic Parts. Vol. 84, pp. 63–84.
Woodward, A. E. and *Sauer, J. A.*: The Dynamik Mechanical Properties of High Polymers at Low Temperatures. Vol. 1, pp. 114–158.
Worsfold, D. J. see Beylen, M. van: Vol. 86, pp. 87–143.
Wunderlich, B.: Crystallization During Polymerization. Vol. 5, pp. 568–619.
Wunderlich, B. and *Baur, H.*: Heat Capacities of Linear High Polymers. Vol. 7, pp. 151–368.
Wunderlich, B. and *Grebowicz, J.*: Thermotropic Mesophases and Mesophase Transitions of Linear, Flexible Macromolecules. Vol. 60/61, pp. 1–60.
Wunderlich, B., Möller, M., Grebowicz, J. and *Baur, H.*: Conformational Motion and Disorder in Low and High Molecular Mass Crystals. Vol. 87, pp. 1–121.
Wrasidlo, W.: Thermal Analysis of Polymers. Vol. 13, pp. 1–99.

Yamashita, Y.: Random and Black Copolymers by Ring-Opening Polymerization. Vol. 28, pp. 1–46.
Yamazaki, N.: Electrolytically Initiated Polymerization. Vol. 6, pp. 377–400.

Yamazaki, N. and *Higashi, F.:* New Condensation Polymerizations by Means of Phosphorus Compounds. Vol. 38, pp. 1–25.
Yilgör, I. and *McGrath, J. E.:* Polysiloxane Containing Copolymers: A Survey of Recent Developments. Vol. 86, pp. 1–86.
Yokoyama, Y. and *Hall, H. K.:* Ring-Opening Polymerization of Atom-Bridged and Bond-Bridged Bicyclic Ethers, Acetals and Orthoesters. Vol. 42, pp. 107–138.
Yorkgitis, E. M., Eiss, N. S. Jr., Tran, C., Wilkes, G. L. and *McGrath, J. E.:* Siloxane-Modified Epoxy Resins. Vol. 72, pp. 79–110.
Yoshida, H. and *Hayashi, K.:* Initiation Process of Radiation-induced Ionic Polymerization as Studied by Electron Spin Resonance. Vol. 6, pp. 401–420.
Young, R. N., Quirk, R. P. and *Fetters, L. J.:* Anionic Polymerization of Non-Polar Monomers Involving Lithium. Vol. 56, pp. 1–90.
Yuki, H. and *Hatada, K.:* Sterospecific Polymerization of Alpha-Substituted Acrylic Acid Esters. Vol. 31, pp. 31, pp. 1–45.

Zachmann, H. G.: Das Kristallisations- und Schmelzverhalten hochpolymerer Stoffe. Vol. 3, pp. 581–687.
Zaikov, G. E. see Aseeva, R. M. Vol. 70, pp. 171–230.
Zakharov, V. A., Bukat, G. D., and *Yermakov, Y. I.:* On the Mechanism of Olifin Polymerization by Ziegler-Natta Catalysts. Vol. 51, pp. 61–100.
Zambelli, A. and *Tosi, C.:* Stereochemistry of Propylene Polymerization. Vol. 15, pp. 31–60.
Zhirkov, P. V. see Malkin, A. Ya.: Vol. 95, pp. 111–148.
Zucchini, U. and *Cecchin, G.:* Control of Molecular-Weight Distribution in Polyolefins Synthesized with Ziegler-Natta Catalytic Systems. Vol. 51, pp. 101–154.
Zweifel, H. see Lohse, F.: Vol. 78, pp. 59–80.

Subject Index

Acetylene derivatives, polycyclotrimerization 50
Acid anhydrides 45, 54
Alternating copolymer 130
Antithrombogenicity 107, 140, 143
Antithrombogenic materials 137
Antitumor activity 107, 123, 126

Base stacking 129, 131
Benzoyl peroxide 54–57
Benzyldimethylamine 48
Benzyltrimethylammonium chloride 52
Bis(4-allyloxyphenyl)sulfone 56
Bis(4-aminophenyl)methane 54
Bis(4-aminophenyl)sulfone 52, 55
9,9-Bis(4-cyanatophenyl)fluorone 46
2,2-Bis[4-(2-hydroxyethoxy)phenyl]propane dimethacrylate 55
Bis(4-hydroxyphenyl)sulfone 52
Bismaleimides 5, 48, 49, 53
Bis(4-maleimidophenyl)ether 48, 49
Bis(4-maleimidophenyl)methane *see* BMI
Blood clotting time 143
–, whole human 143
BMI 48, 49, 52–56
BPA bis(vinylbenzyl)ether 55
BPA/DC-BMI prepolymer 54, 55
BPA/DC prepolymer 44, 48, 52, 54, 57
BPA diglycidyl ether 52
BPA monocyanate 45, 52, 56
BT resins 49
tert Butyl peroxide 54, 56

Carbon fiber composites 45, 55
Carborane 44
Carboxylfluorecein release 133
CF-release 133
Charge transfer (CT) complex 9, 10, 11, 21, 23
– – polymerization 3
Chelates, metal 49
–, polymeric 64
Chelating ion-exchange resins 64
– macroligand systems 65f.
– polymers, crosslinked 95
Co naphthenate 52

Conductive materials 52
Copper clad laminates 48, 49, 52, 53, 55–57
– naphthenate 45
– wire enamelling 55
Cumyl cyanate 43, 52
Cumylphenyl cyanate 49
Cyanate/epoxide/maleimide compositions 53, 54
Cyanate/maleimide compositions 48, 49, 52
Cyanates, cyclotrimerization 44
1-(2-Cyanoethyl)-2-phenylimidazole 51, 52
Cyanurate rings 46
Cycloaddition 4, 8, 14, 16, 26, 31
Cyclotrimerization of cyanates, activation energy 44

DABCO *see* Triethylenediamine
Derivatography 56
Diallyl phthalate 56
N-(3,4-Dichlorophenyl)-N',N'-dimethylurea 55
Dicumyl peroxide 45, 54, 56
Dicyanate/epoxide compositions 49–53
Dicyanate polyaddition 46
Dicyandiamide 52
Dicyclopentadiene 54
Dielectronic properties 57
Diisocyanates 46
4,4-Diisocyanatodiphenylmethane (MDI) 57
Dimerization 16–18
– cyclodimerization 9
Dimethylaniline 56
Divinylbenzene 55
Drug delivery system 132
Drug-immobilized biomaterials 107
Drug release 107
– – by hydrolysis 116
Drugs, polymers 107ff.

ECH *see* Epichlorohydrin
EDA complex 10
Ehrlich's ascites 125, 126
– – tumor-bearing mice 123
Electroinsulating coating 57
Electron transfer 17, 18, 20
Epichlorohydrin 50, 52, 55, 56

Epoxide/cyanate cycloaddition 53
Epoxide/maleimide systems 53
Epoxide resins 49–54
Epoxyacrylate resins see Vinyl ester resin
Epoxynovolak resin 50, 52, 54, 55
— —, brominated 54
Ethyl acetoacetate 57
Ethylene-propylene rubber 57
2-Ethylhexyl acrylate 48
2-Ethylimidazole 52
2-Ethyl-4-methylimidazole 54, 55
Exciplex 4, 9, 11, 16, 25, 26, 31, 32
— polarity 14
— singlet 14, 29
— triplet 14, 16
Excited complex 11
Equilibrium constant 10

Fatty acids, dimerized 57
Flory's principle 67
Fluorescence 13, 16
5-Fluorouracil 107ff.
5-Fluorouracil, N-substituted 110
5-Fluorouracil, vinyl monomers 108

Glycidyl methacrylate 48, 55

Heat conductive molds 45
Heparin 107, 137
— release 139, 140
Heterometallic complexes, immobilized 92
Hexamethylene tetraamine 57
Human neutrophils 134, 135
Hydrogels 137, 139
Hydrogen bonding 129, 131
Hydrolysis 107
— of polymeric drugs 116ff.
2-Hydroxyethyl methacrylate 48
Hypochromicity 128, 129, 131, 132

Imidocarbamates 45
Imidocarbonates 45
Immunomodulating activity 134
— — of polymer drugs 107
Immonomodulators, polymer 136
Impregnation 55
Initiating benzophenone/aniline 4
— systems 3
Injection molding powders 49
Interpenetrating polymer networks see IPN
Ion-radical 13, 16, 17, 22, 31
IPN 47, 48, 52, 55, 56
—, simultaneous 47, 56
Irradiation 9, 11
—, CT complex 9
—, monomers 9
—, triplet sensitizer 9, 14

IR spectra 44, 56
Isocyanurate rings 46
p-Isopropenylphenyl cyanate 48
p-Isopropylphenyl cyanate see Cumyl cyanate

Ladder polymers 48
Liposome 107, 132, 133, 134, 135

Macroligands, chelating 65f, 77
—, crosslinked 95
Macrophage activation 133
Macrophages, mouse peritoneal 135, 136
Maleimide/cyanate cycloaddition 53
Maleimides 47, 48
Metal chelates 49
Metallocyclization 64
Metallochelate-containing soluble polymers 93f.
— polymers 64
— —, polynuclear 90
— units, polymers containing (PCMU) 61ff.
Metallochelates, catalytic activity 99
— macromolecular 99
Metallomonomers 83
Metallopolymeric systems 61ff.
Methacrylic acid 56
Methacryloyloxyethyl-type monomers 111
Microspheres 113, 120
Molds, heat conductive 45
Monomer excitation 11, 13
Mouse macrophage 135, 136

Neutrophils, human 134, 135
Nitrile rubber 57
Nonylphenyl 45
Norrish Type II 24
Nucleic acid analogs 107, 127, 129
— — —, water-soluble 133

Oligoaspartimides 49, 53, 54
Oligodimethylsiloxane 44
Oligomethylphenylsiloxane 44
Oxazoline 50, 53

Paterno-Buchi reaction 25
PCMU — Polymers containing metallochelate units 61ff.
Phenolic resins 57
Phenyl glycidyl ether 49
2-Phenylimidazole 54
Photo-induced copolymerization 1ff, 9, 12
Photopolymerization 1ff, 3, 5, 7ff, 13
—, acrylonitrile 6, 8
—, fumaronitrile 8
—, initiators 9
—, maleicanhydride 7

Subject Index

Photopolymerization, α-methylstyrene 5
—, N-vinylcarbazole 5
Platelet aggregation 141
— —, inhibition 142
Polyacrylate 54
Polyanionic polymers 107, 123, 125, 127, 132, 133, 136
Polybutadiene 48, 52, 55
Polycarbonate 47
Polychelate effects 68, 75
Polycyanurates 43, 45
Polycyclotrimerization of acetylene derivatives 50
— of dicyanates 43, 44, 50, 53
Poly(2,6-dimethylphenylene oxide) see PPO
Polyesters 49, 52, 55, 57
Polyetherimide 49
Polyethersulfone 47, 54, 55
Polyhydantoin 55
Polyimide 48
Polyimidocarbamates 45
Polyiminocarbonates 46
Polyisoureas 46, 54
Polymer-analogous reactions 82
Polymer drugs 107ff.
— —, water-soluble 112, 133
— —, synthesis 111
—, intercalating 132
Polymeric chelates 64
Polynuclear metallochelate polymers 90f.
Polyoxypropylene 47, 57
Poly(phenylene sulfide) 52
Polyphenylquinoxaline 48
Polypropylene 47
Polystyrene 48
Polysulfone 52, 54
Polytetrafluoroethylene 53
Polytriazines 43
Polyurethanes 57
Poly(vinyl chloride) (PVC) 137, 139
Powder coatings 49
PPO 47, 48, 56
Preimpregnates 49
Prodrugs, polymeric 107f.
Prostaglandins 107, 137
Proton transfer 6, 20
PVC 137, 139
Pyromellitic dianhydride 45, 51, 52, 54

Resol 57
RNA 129, 132
Rubber 57

Sarcoma 180 ascites 124, 126
Segmental miscibility 56

Semi-IPN 47, 55
Semi-ladder polymers 48
Solvent effect 9, 20
Solvolysis 107, 110
Spontaneous polymerization 13, 21
Step-growth mechanism 44
Styrene 48, 56
Sulfur 57
Superoxide anion production 134, 135, 136

Tetrabromo-BPA 52, 54
Tetracarboxylic acid dianhydrides 45
Tetraepoxide 52, 54
Tetraethylammonium bromide 51
Tetraethylenepentaamine 51
Tetramethylene, diradical 1ff, 21, 23, 28, 29, 32
—, — in cycloadditions 26, 30, 33
—, — photodimerization 29
—, — singlet 24, 25, 30
—, — triplet 24, 35
— intermediates 17, 21
—, photochemistry 23
—, thermal copolymerization 21
—, zwitterionic 21, 22, 28, 30
Theophylline 110
Thermal decomposition 45
Thermomechanical curve 56
Thermooxidative stability 50
Thermoplastic polymers 47
Thermostable polymers 98
p-Toluenesulfonic acid 45, 49
p-Toluenesulfonic monohydrate 45, 55
Tokal packed cell volume 123, 125
Triallyl isocyanurate (TAIC) 56
Triazine A resin 45
— polymers 43
Triethylenediamine 49, 52, 54–57
Triglycidyl-p-aminophenol 52
Trimellitic anhydride 55, 57
Trimethylene bis(4-aminobenzoate) 56
Trimethylolpropane trimethacrylate 55
Triplex 15
1,1,1-Tris(4-cyanatophenyl)ethane 45

UV irradiation 48

9-**V**inyladenine 127
Vinyl ester resin 56
— imidazoles 117
— monomers having 5-fluorouracil 108

Zn acetate 45, 49, 52, 54-56
Zn naphthenate 45
Zn octoate 48, 49, 52, 55-57